Selected Titles in This Series

621 **Gregory L. Cherlin,** The classification of countable homogeneous directed graphs and countable homogeneous n-tournaments, 1998

620 **Victor Guba and Mark Sapir,** Diagram groups, 1997

619 **Kazuyoshi Kiyohara,** Two classes of Riemannian manifolds whose geodesic flows are integrable, 1997

618 **Karl H. Hofmann and Wolfgang A. F. Ruppert,** Lie groups and subsemigroups with surjective exponential function, 1997

617 **Robin Hartshorne,** Families of curves in $\mathbb{P}^3$ and Zeuthen's problem, 1997

616 **Serguei G. Bobkov and Christian Houdré,** Some connections between isoperimetric and Sobolev-type inequalities, 1997

615 **Michael A. Dritschel and Hugo J. Woerdeman,** Model theory and linear extreme points in the numerical radius unit ball, 1997

614 **Richard Warren,** The structure of k-CS-transitive cycle-free partial orders, 1997

613 **D. L. Flannery,** The finite irreducible linear 2-groups of degree 4, 1997

612 **Joan Porti,** Torsion de Reidemeister pour les variétés hyperboliques, 1997

611 **D. Ginzburg, I. Piatetski-Shapiro, and S. Rallis,** L functions for the orthogonal group, 1997

610 **Mark Hovey, John H. Palmieri, and Neil P. Strickland,** Axiomatic stable homotopy theory, 1997

609 **Liviu I. Nicolaescu,** Generalized symplectic geometries and the index of families of elliptic problems, 1997

608 **Christina Q. He and Michel L. Lapidus,** Generalized Minkowski content, spectrum of fractal drums, fractal strings, and the Riemann zeta-functions, 1997

607 **Adele Zucchi,** Operators of class C_0 with spectra in multiply connected regions, 1997

606 **Moshé Flato, Jacques C. H. Simon, and Erik Taflin,** Asymptotic completeness, global existence and the infrared problem for the Maxwell-Dirac equations, 1997

605 **Liangqing Li,** Classification of simple C^*-algebras: Inductive limits of matrix algebras over trees, 1997

604 **Hajnal Andréka, Steven Givant, and István Németi,** Decision problems for equational theories of relation algebras, 1997

603 **Bruce N. Allison, Saeid Azam, Stephen Berman, Yun Gao, and Arturo Pianzola,** Extended affine Lie algebras and their root systems, 1997

602 **Igor Fulman,** Crossed products of von Neumann algebras by equivalence relations and their subalgebras, 1997

601 **Jack E. Graver and Mark E. Watkins,** Locally finite, planar, edge-transitive graphs, 1997

600 **Ambar Sengupta,** Gauge theory on compact surfaces, 1997

599 **Tai-Ping Liu and Yanni Zeng,** Large time behavior of solutions for general quasilinear hyperbolic-parabolic systems of conservation laws, 1997

598 **Valentina Barucci, David E. Dobbs, and Marco Fontana,** Maximality properties in numerical semigroups and applications to one-dimensional analytically irreducible local domains, 1997

597 **Ragnar-Olaf Buchweitz and John J. Millson,** CR-geometry and deformations of isolated singularities, 1997

596 **Paul S. Bourdon and Joel H. Shapiro,** Cyclic phenomena for composition operators, 1997

595 **Eldar Straume,** Compact connected Lie transformation groups on spheres with low cohomogeneity, II, 1997

594 **Solomon Friedberg and Hervé Jacquet,** The fundamental lemma for the Shalika subgroup of $GL(4)$, 1996

593 **Ajit Iqbal Singh,** Completely positive hypergroup actions, 1996

(*Continued in the back of this publication*)

The Classification of Countable Homogeneous Directed Graphs and Countable Homogeneous n–tournaments

Memoirs
of the
American Mathematical Society

Number 621

The Classification of Countable Homogeneous Directed Graphs and Countable Homogeneous n–tournaments

Gregory L. Cherlin

January 1998 • Volume 131 • Number 621 (first of 4 numbers) • ISSN 0065-9266

American Mathematical Society
Providence, Rhode Island

1991 *Mathematics Subject Classification.*
Primary 05C20; Secondary 03C10, 03C15, 03C50, 05D10, 20B22, 20B27.

The photograph on the opposite page is courtesy of Frans Lanting/Minden Pictures.

Library of Congress Cataloging-in-Publication Data

Cherlin, Gregory L., 1948–
  The classification of countable homogeneous directed graphs and countable homogeneous n-tournaments / Gregory L. Cherlin.
    p. cm. — (Memoirs of the American Mathematical Society, ISSN 0065-9266 ; no. 621)
  "January 1998, volume 131, number 621 (first of 4 numbers)."
  Includes bibliographical references and indexes.
  ISBN 0-8218-0836-2 (alk. paper)
    1. Directed graphs. 2. Tournaments (Graph theory) 3. Model theory. 4. Ramsey theory. 5. Permutation groups. I. Title. II. Series.
QA3.A57 no. 621
[QA166.15]
510—dc21
[511'.5]   97-31683
                                                                             CIP

Memoirs of the American Mathematical Society

This journal is devoted entirely to research in pure and applied mathematics.

Subscription information. The 1998 subscription begins with volume 131 and consists of six mailings, each containing one or more numbers. Subscription prices for 1998 are $435 list, $348 institutional member. A late charge of 10% of the subscription price will be imposed on orders received from nonmembers after January 1 of the subscription year. Subscribers outside the United States and India must pay a postage surcharge of $30; subscribers in India must pay a postage surcharge of $43. Expedited delivery to destinations in North America $35; elsewhere $110. Each number may be ordered separately; *please specify number* when ordering an individual number. For prices and titles of recently released numbers, see the New Publications sections of the *Notices of the American Mathematical Society.*

Back number information. For back issues see the *AMS Catalog of Publications.*

Subscriptions and orders should be addressed to the American Mathematical Society, P. O. Box 5904, Boston, MA 02206-5904. *All orders must be accompanied by payment.* Other correspondence should be addressed to Box 6248, Providence, RI 02940-6248.

Copying and reprinting. Individual readers of this publication, and nonprofit libraries acting for them, are permitted to make fair use of the material, such as to copy a chapter for use in teaching or research. Permission is granted to quote brief passages from this publication in reviews, provided the customary acknowledgment of the source is given.

Republication, systematic copying, or multiple reproduction of any material in this publication (including abstracts) is permitted only under license from the American Mathematical Society. Requests for such permission should be addressed to the Assistant to the Publisher, American Mathematical Society, P. O. Box 6248, Providence, Rhode Island 02940-6248. Requests can also be made by e-mail to `reprint-permission@ams.org`.

Memoirs of the American Mathematical Society is published bimonthly (each volume consisting usually of more than one number) by the American Mathematical Society at 201 Charles Street, Providence, RI 02904-2294. Periodicals postage paid at Providence, RI. Postmaster: Send address changes to Memoirs, American Mathematical Society, P. O. Box 6248, Providence, RI 02940-6248.

© 1998 by the American Mathematical Society. All rights reserved.
This publication is indexed in *Science Citation Index*®, *SciSearch*®, *Research Alert*®, *CompuMath Citation Index*®, *Current Contents*®/*Physical, Chemical & Earth Sciences.*
Printed in the United States of America.

∞ The paper used in this book is acid-free and falls within the guidelines
established to ensure permanence and durability.
Visit the AMS home page at URL: http://www.ams.org/

10 9 8 7 6 5 4 3 2 1   03 02 01 00 99 98

Dedicated to the memory of Alan Mekler

1947 - 1992

It's guid to be merry and wise.

- Robert Burns

Contents

Introduction — xi

Chapter 1. Results and Open Problems — 1
  1.1. Homogeneous structures. — 1
  1.2. A survey of work on homogeneous structures. — 3
  1.3. Amalgamation classes. — 6
  1.4. Languages, strong amalgamation, generification, and Ramsey's theorem — 9
  1.5. Classification theorems. — 12
  1.6. Open problems. — 13

Chapter 2. Homogeneous 2-tournaments — 17
  2.1. A catalog. — 17
  2.2. Restricted homogeneous 2-tournaments. — 21
  2.3. Sources and sinks — 26
  2.4. Constrained 2-tournaments — 28
  2.5. Unconstrained 2-tournaments — 33

Chapter 3. Homogeneous n-tournaments — 38
  3.1. Introduction — 38
  3.2. Hypercritical and small 3-tournaments — 39
  3.3. The critical case — 40
  3.4. Two embedding lemmas. — 45
  3.5. Polarized n-tournaments — 47
  3.6. Embedding polarized 3-tournaments — 48
  3.7. Some special cases — 49
  3.8. The general case — 51

Chapter 4. Homogeneous symmetric graphs — 53
  4.1. The theorem of Lachlan and Woodrow — 53
  4.2. The main ingredients — 55
  4.3. Structure of the proof — 56
  4.4. Steps 7, 5, 8. Proof of the Main Theorems — 56
  4.5. Step 1, Proposition 10: adding $K(2)$ — 58
  4.6. Step 1, Proposition 11: the operation H^+ — 60
  4.7. Step 1, Propositions 12 and 13: realization of 1-types — 62
  4.8. Step 2. Theorem 4.8: a, b, K — 66
  4.9. Step 6. Theorem 4.9.n: extending direct sums — 66
  4.10. Step 3. Theorem 4.6 — 66
  4.11. Step 4. Theorem 4.7 — 67

Chapter 5. Homogeneous directed graphs omitting I_∞ — 74
- 5.1. A catalog of homogeneous directed graphs — 74
- 5.2. The graph $\mathcal{P}(3)$ — 76
- 5.3. The theorem — 77
- 5.4. The major steps in the proof — 79
- 5.5. Proof of the Main Theorem, Part 1 — 80
- 5.6. Proof of the Main Theorem, part 2, $n > 2$ — 82
- 5.7. Case 2.1 of the Main Theorem — 83
- 5.8. Propositions 14 and 15 — 84

Chapter 6. Propositions 16 to 20 and MT 2.2 — 91
- 6.1. Proposition 16: simple configurations — 91
- 6.2. Proposition 17: induction on n — 93
- 6.3. Proposition 18: extending I_n — 95
- 6.4. Proposition 19: $(^py, y')$ — 96
- 6.5. Proposition 20: $(^py, y^\perp)$ — 97
- 6.6. Toward MT 2.2 — 99
- 6.7. Lemma 6.3 — 101
- 6.8. Lemma 6.4 — 104
- 6.9. Lemma 6.5 — 110
- 6.10. Lemma 6.6 — 117

Chapter 7. Homogeneous directed graphs embedding I_∞ — 119
- 7.1. The classification theorem — 119
- 7.2. The main ingredients — 121
- 7.3. Structure of the proof — 122
- 7.4. Steps 4, 6, 7. The Main Theorem — 123
- 7.5. Step 1. Proposition 24: P_3 — 125
- 7.6. Step 1, Proposition 25: adding $L(2)$ — 128
- 7.7. Step 1, Proposition 26: the operations $\pm$ — 131
- 7.8. Step 1, Propositions 27 and 28: some 1-types — 133

Chapter 8. Theorems 7.6-7.9 — 138
- 8.1. Step 2. Theorems 7.6 and 7.7 — 138
- 8.2. Step 5. Theorem 7.9.$\mathcal{T}$: extending a direct sum — 141
- 8.3. Step 3. Theorem 7.8, 1-types over sums — 141
- 8.4. Theorem 7.8, conclusion — 144

Appendix: Examples for richer languages — 150

Bibliography — 154

Index of Notation — 158

Index — 159

ABSTRACT. The main new result given here is the classification of the countable homogeneous directed graphs, carried out in chapters 5-8. It has long been known that there are $2^{\aleph_0}$ such graphs. The classification of homogeneous n-tournaments given in chapters 2-3 serves to illustrate the methods and is occasionally useful in the latter half of the monograph. Chapter 4 gives a new proof of the corresponding theorem of Lachlan and Woodrow for symmetric graphs, in a manner quite similar to that used here in the directed case. An appendix points out some examples of homogeneous structures in richer languages which one would probably want to consider more closely before attempting to carry out arguments of a similar type in richer binary languages.

1991 *Mathematics Subject Classification*. Primary 05C20;
Secondary 03C10, 03C15, 03C50, 05D10, 20B22, 20B27.
Keywords: homogeneity, digraph, ramsey, tournament, graph, automorphism, amalgamation, model theory, classification.

Introduction

When combinatorial methods emerged which proved adequate for the classification of all countable homogeneous graphs [61] and all countable homogeneous tournaments [57], I thought it might be enlightening to tackle the case of countable homogeneous directed graphs, since there were $2^{\aleph_0}$ examples already known in that case [42], and there is a natural tendency to take uncountability as evidence for the intrinsic unclassifiability of such classes of structures. This project eventually succeeded. The resulting classification is the main result of the present Memoir. It is given as a catalog in Chapter 5, and is proved in Chapters 5-8. Judging by the length of the arguments presented there, this may take us close to the limits of usefulness of these methods – but the jury is still out.

For a discussion of the general classification problem for homogeneous structures, see Lachlan's report to the 1986 ICM in [59]. I have provided an introduction to the subject in Chapter 1, together with some discussion of related open questions. Subsequent chapters contain the details of the proofs of the two results described in the abstract. The classification results for homogeneous structures presented in Chapters 2-3 (n-tournaments), 5-6 (directed graphs omitting I_∞), and 7-8 (directed graphs embedding I_∞) can be read independently of one another, though I have taken advantage of the Memoir format to absorb most of the background material into Chapter 1. Chapter 4 was added later, to show how the proof for directed graphs would go in the case of symmetric graphs. This proof deviates in interesting ways from the proof given in [61] for the symmetric case.

There are a number of other classification problems for homogeneous structures for which it seems unreasonable to expect an explicit classification, notably the classification of homogeneous nil rings and homogeneous nilpotent groups; but the only hard evidence to support that belief lies in the construction of $2^{\aleph_0}$ examples by the *same methods* that produce $2^{\aleph_0}$ directed graphs. Consequently the classification of the homogeneous directed graphs tends to underscore the extent to which the difficulties that arise in these algebraic situations are still not properly understood.

I wish to record my gratitude to Cheryl Chute Miller, Alistair Lachlan, and Carol Wood, as well as a singularly attentive and industrious referee, for their assistance in making some very lengthy arguments clearer and more accurate than they would otherwise be. Not all of their suggestions have been adopted, for which the author naturally bears full blame.

Gregory Cherlin

CHAPTER 1

Results and Open Problems

1.1. Homogeneous structures.

A structure Γ is said to be *homogeneous* (in the sense of Fraïssé [**35**]) if any isomorphism $\iota : A \simeq B$ between any two of its finitely generated substructures A, B is induced by some automorphism of Γ; in other words, Aut Γ acts as transitively as possible, given that Γ already carries some structure. Homogeneous structures have been studied from a number of different points of view, by logicians, algebraists, and combinatorialists, with a view either toward their classification or toward a better understanding of individual examples. In particular considerable attention has been paid to the intricate structure of their automorphism groups, either as permutation groups or as abstract groups.

The present Memoir is devoted to two quite specific classification theorems. We give an explicit description of all the countable homogeneous "n-tournaments" (n-tournaments are a natural generalization of n-vertex colored tournaments), and all the countable homogeneous directed graphs. Our techniques derive entirely from one paper by Lachlan [**57**], which at present provides the most powerful approach to the classification of infinite homogeneous structures of combinatorial (as opposed to algebraic) type; this distinction, which is not intended to be precise, can be taken to refer to the presence or absence of functions (as opposed to relations) in the structure Γ. In the absence of functions, any subset of Γ is the domain of a substructure, and the homogeneity condition is then extremely powerful.

It will be seen that these two results, like other classification theorems, require a surprisingly lengthy argument, even though the classifications themselves are not very complex. It is not at all clear as yet why this is the case. The most obvious possibility is that our combinatorial tools need to be refined. A more intriguing possibility is that the classification of more general structures cannot be carried out effectively. This effectivity problem can be posed quite precisely, and is discussed in some detail in [**59**]. We shall return to this question below after reviewing the machinery of amalgamation classes.

The classification of the countable homogeneous directed graphs was the main goal of the present enterprise, because it has been known for about twenty years that there are $2^{\aleph_0}$ examples, and this has tended to cast a pall over the whole idea of classification in this context. In fact, the present classification was not undertaken in the expectation that it would actually be completed, but rather with the feeling that it would be interesting to see how it breaks down. Instead, we arrived at a result which may be phrased as follows: *the proof in* [**42**] *that there are $2^{\aleph_0}$ homogeneous directed graphs actually contains a simple construction procedure which produces all the homogeneous directed graphs except for a countable number of exceptional graphs, which fall naturally into finitely many parametrized families.*

The classification of the countable homogeneous n-tournaments is considerably easier than the classification of the homogeneous directed graphs, but it also involves a point of some methodological interest. The methods we use here seem limited in their effect to the classification of homogeneous binary structures, that is, structures equipped with unary predicates (a vertex coloring) and binary relations exclusively, and these methods are also extremely sensitive to the number of binary relations present. Previous classification theorems for homogeneous structures only covered finitely many types of binary structures, apart from degenerate cases. The class of n-tournaments, as defined in §4 below, involves an unlimited but finite supply of binary relations, even with n fixed. (The number of binary relations is one of the parameters that winds up being fed into an inductive argument.) It turns out that there are all together only countably many countable homogeneous n-tournaments (which we describe explicitly), so in this sense the class of n-tournaments is less complex than the class of directed graphs.

At present it is very difficult to see whether the sort of result proved here is just the tip of the iceberg, or whether we have reached a point at which the existing methods begin to collapse under their own weight. (Some may feel we have passed this point.) The natural open problems arising in the continuation of this work will be discussed at the end of this chapter.

The present chapter is devoted to a survey of background material. We begin with a general survey of a some earlier work on homogeneous structures. None of this is in any way essential to an understanding of the work reported here. In the remainder of the chapter we review Fraïssé's theory of amalgamation classes, which is an essential ingredient in our work, and we summarize our results, leaving a more detailed presentation, including an explicit description of all the examples, to the beginning of Chapters 2 and 5. The chapter concludes with a discussion of some open problems.

In Chapters 2,3 we carry out the classification of homogeneous n-tournaments. Chapters 5-8 contain the classification of homogeneous directed graphs, modulo the results of [**56, 17**] which cover the finite and imprimitive cases. (A structure Γ is said to be primitive if its automorphism group is primitive, that is, $\mathrm{Aut}\,\Gamma$ leaves no non-trivial equivalence relation invariant.) These four chapters consist mainly of two long inductive proofs, each one spread out over two chapters. We first treat the classification of the primitive homogeneous directed graphs which do not contain an infinite edgeless induced graph, and then those which do; as it turns out, this case division supplies a very useful hypothesis in each of the two cases.

The intermediate Chapter 4 is devoted to a new proof of the Lachlan/Woodrow classification of the homogeneous undirected graphs, along the lines used in Chapters 7-8 in connection with directed graphs. This proof is included here as a simpler illustration of methods used in the directed case. It was obtained in a roundabout manner: the methods used to classify homogeneous tournaments were extended to the case of directed graphs, then specialized back to the undirected case. The part of the classification of the homogeneous directed graphs given in Chapters 5-6 parallels the analysis of homogeneous tournaments particularly closely, and the rest of the analysis in Chapters 7-8 is a natural generalization of the argument given in Chapter 4. Chapter 4 can be omitted in its entirety without loss.

We note that all structures considered seriously here are countable, and occasionally finite. It seems worthwhile to continue to mention the countability hypothesis explicitly until we complete our review of Fraïssé's theory in §3 below, after which this assumption will usually be left tacit. Set theoretical issues intervene massively in the classification of homogeneous structures of uncountable cardinality, but every such structure does give rise in a canonical way to a countable homogeneous structure of a very similar type; this follows from elementary results in model theory, and is clear in any case on the basis of the theory presented in §3.

1.2. A survey of work on homogeneous structures.

It should be said at the outset that the survey below is primarily a review of *model theoretic* work on homogeneous structures. Algebraic work tends to focus more on the structure of the automorphism group or on numerical properties of associated permutation representations of the automorphism group, though there is no hard and fast division between the two points of view.

The archetypal examples of homogeneous structures are vector spaces over arbitrary fields, and countable dense linear orderings. The dichotomy between highly structured, algebraic examples, and comparatively amorphous, combinatorial structures has persisted in later analysis, though the line between the two situations becomes blurred; finite examples tend to be highly structured, whatever their origin, while infinite locally finite algebraic systems can resemble their more amorphous counterparts.

The general notion of homogeneity was studied by Roland Fraïssé in [31, 35, 33, 32], where the connection with the amalgamation property (reviewed in the next section) was uncovered. Around 1970 it was realized that Fraïssé's idea provided an easy approach to the construction of a bewildering variety of infinite structures Γ with a highly transitive automorphism group, (that is, with $\Gamma^n/\operatorname{Aut}\Gamma$ finite for every n). In particular one can easily construct nontrivial n-transitive groups for arbitrary values of n by this method, and at the same time retain precise control over the degree of complexity of the first order theory of the resulting structure.

In the mid-seventies the paper [66] sparked an interest in the study of the classification of first order theories of algebraic systems admitting quantifier elimination in their natural languages. A discussion of quantifier elimination would take us too far afield; the important point here is that eliminability of quantifiers and homogeneity are closely related, and in certain fairly broad contexts are even equivalent. The work in [66] is not in fact concerned with homogeneity, but as this work was extended it naturally led to a number of concrete classification problems for homogeneous rings, [6, 3, 4, 5]. In fact the classification of rings admitting quantifier elimination was reduced fairly rapidly to the classification of countable homogeneous uniformly locally finite rings. (The local finiteness means of course that finitely generated subrings are finite, and the uniformity means that the number of elements in a finitely generated subring is bounded by a function of the number of generators.) This problem was solved completely in the finite case, [78, 81, 79, 80] but there are $2^{\aleph_0}$ countable homogeneous nil rings (even in the commutative case) [5, 78], for much the same reason as in the case of directed graphs: Fraïssé's technology. Analogous work on groups with quantifier elimination and homogeneous groups led to quite similar results, in a more limited context: the finite and solvable cases are understood [21, 20, 69], modulo difficulties in the specific case of nilpotent groups of class 2 and exponent 4; and there are again $2^{\aleph_0}$

homogeneous groups of this specific type, for the usual reasons [**77**] – though with striking technical complications.

A quite different line of investigation developed about the same time. J. Sheehan [**83**], A. Gardiner [**36**], and Klin and Golfand [**39**], independently classified the finite homogeneous undirected graphs. There are very few: apart from the pentagon C_5 and one nine-vertex graph $E(K_{3,3})$ which may be described as the edge graph of the complete bipartite graph $K_{3,3}$, every other homogeneous graph is of the form $m \cdot K_n$, the disjoint union of m copies of the complete graph on n vertices, or its complement, the complete m-partite graph with n vertices in each component of the partition. Lachlan and Woodrow [**61**] completed the classification of the countable homogeneous graphs by classifying all of the infinite ones, using a very ingenious inductive argument, and bringing Fraïssé's theory to bear. This seems to be the first occasion on which Fraïssé's theory was used to *limit* the possible homogeneous structures. In addition to the natural infinite analogs of the finite homogeneous graphs, there is a single infinite series $\{\Gamma_n : n \leq \infty\}$ of examples which have no finite analogs. Γ_∞ is Rado's graph, also known as the random countable graph, and Γ_n is an analogous graph omitting the complete graph K_{n+1}. The graphs $C_5, E(K_{3,3}), m \cdot K_n$ and Γ_n $(m, n \leq \infty)$, together with the complementary graphs, constitute all the countable homogeneous graphs.

Lachlan then embarked on a massive generalization of some of the qualitative aspects of this result to a much broader context: the classification of the finite homogeneous structures for an arbitrary finite relational language, [**58, 60, 22, 51**] and more generally still, the homogeneous structures (for finite relational languages) which are stable in Shelah's sense. Lachlan showed that all finite or stable structures which are homogeneous for a fixed, finite, relational language, can be divided into finitely many families, with the structures in each family naturally parametrized by numerical invariants, typified by the parameters m, n in the case of the family $m \cdot K_n$. For a technical reason the theory was not quite satisfactory in the original formulation in [**58**]; it was based on an explicit additional assumption regarding a notion of "rank", formulated as a conjecture in [**58**]. This conjecture was proved by a direct argument in [**60**] for the case of *binary* languages, and by an argument based on the classification of the finite simple groups in [**22**], which works for arbitrary finite relational languages; the group theoretic portion of this latter analysis has since been pushed much further [**49**].

An immediate consequence of Lachlan's theory is that there are only countably many countable homogeneous structures for a finite relational language which are stable in Shelah's sense. Lachlan's theory describes roughly half of the picture that is visible in the case of homogeneous graphs, and allows us to identify the class consisting of the finite graphs and their natural infinite analogs with a class arising independently in model theory.

Lachlan subsequently classified the countable homogeneous tournaments [**57**] by combining the sort of induction which had surfaced in [**61**] with an additional argument based on Ramsey's theorem; these two ideas will be used repeatedly in the present Memoir. In the process of working out the case of directed graphs, it became clear that these ideas could be used somewhat more efficiently in the case of tournaments, without making any essential modifications; this more efficient presentation is found in [**18**], and corresponds fairly closely in its overall structure to the more elaborate arguments given here, especially in Chapter 6. As it turns out, there are only five countable homogeneous tournaments, and the bulk of the

1.2. A SURVEY OF WORK ON HOMOGENEOUS STRUCTURES.

analysis in [57] or [18] is concerned with providing a characterization of one of them, the natural analog for tournaments of the Rado graph. Nonetheless, the techniques developed for this single case are equally useful in the case of homogeneous directed graphs.

These are the main threads of the story bearing on the work presented here. We shall look more briefly at some other aspects of the study of homogeneous structures.

Some problems that arose originally in the specific case of the Rado graph have led to problems in the theory of homogeneous relational structures. The most outstanding example of this is the Nešetril-Rödl theorem, or family of theorems, generalizing the Ramsey theorem. In one special case, it states that if any finite graph G is specified as a "target", then there is a finite graph G^* which has the property that for any coloring of the edges of G^* by a fixed number of colors, there is a subgraph of G^* which is isomorphic to G, and which is monochromatic in the sense that the edges are all of one color. This has been generalized to a broader class of homogeneous structures [70].

A very striking consequence of the Nešetril-Rödl theorem was found by Simon Thomas [87]. Let Γ be the Rado graph, and view $\operatorname{Aut}\Gamma$ as a subgroup of the symmetric group on the vertices of Γ. This group carries a natural topology, as a subspace of Γ^Γ in the product topology, with Γ discrete: this is the topology of pointwise convergence. The closed subgroups of $\operatorname{Sym}(\Gamma)$ containing $\operatorname{Aut}\Gamma$ are called *reducts* of Γ, or of $\operatorname{Aut}\Gamma$, for model theoretic reasons. Thomas has shown that $\operatorname{Aut}\Gamma$ has only four nontrivial reducts (counting $\operatorname{Aut}\Gamma$ itself, but not counting $\operatorname{Sym}(\Gamma)$). Apart from an ingenious case division corresponding nicely to the structure of the reducts, the main ingredient is the Nešetril-Rödl theorem. This work has been extended by Thomas' student, James Bennett.

While the Rado graph provides a very natural example of a structure which is homogeneous for a finite relational language, an even more natural example is provided by a countable set S carrying no structure at all, apart from the relation of equality. In fact the associated automorphism group, $\operatorname{Sym} S$, is a fairly complicated object with nontrivial properties. The lattice of normal subgroups was identified in [1], and this sort of work has been extended to the Rado graph and similar objects by Truss [89, 90, 91] as well as to linear analogs [74]. A curious property emerged in [37, 24]: any subgroup of $\operatorname{Sym} S$ of index less than $2^{\aleph_0}$ contains the pointwise stabilizer of a finite subset. This property is referred to as the property of "no small index", and has been intensively investigated in various concrete contexts. Thus it holds for the infinite classical groups over finite fields [29] and for the rational order [91], and the generalization to uncountable cardinalities leads to set-theoretic issues like the existence of measurable cardinals [84]. From a model theoretic point of view, this result is particularly striking, because whenever Γ is a countable structure such that $\operatorname{Aut}\Gamma$ has the property of no small index, then for any structure Δ with $\operatorname{Aut}\Delta \simeq \operatorname{Aut}\Gamma$, assuming that $\operatorname{Aut}\Delta$ has only finitely many orbits on Δ, the structure Δ is interpretable within Γ, and in a number of fairly common situations this can force $\Delta \simeq \Gamma$. The more general question, to what extent the abstract group $\operatorname{Aut}\Gamma$ carries information about Γ (assuming that Γ is primitive and under various supplementary conditions) is of interest generally from a model theoretic point of view. The question boils down essentially to whether the natural topology on $\operatorname{Aut}\Gamma$ induced by its action on Γ can be recovered from the abstract structure of the group $\operatorname{Aut}\Gamma$. This has been intensively studied by Mati Rubin in a number of papers, most notably [**75**]. Rubin's results allow one to recover the topology on certain automorphism groups from the abstract group structure without proving the no-small-index property.

In a more numerical vein, it has turned out to be quite interesting to study the orbits of Aut Γ on *unordered n-tuples* from Γ. It was shown in [**7**] that if s_n is the number of such orbits, then the sequence (s_n) is monotonically nondecreasing (see also [**65, 48, 9**]). In the linear analog, Γ carries a vector space structure and s_n counts the number of orbits of Aut Γ on *subspaces*. These numbers are also nondecreasing [**86**]. Accordingly the cases in which a nontrivial equality occurs ($s_n = s_{n+1} > 1$) are extremal, and have been investigated in [**10, 12, 11, 15, 23**].

Finally, we may mention that Hrushovski has developed the Fraïssé construction in new directions, providing a far more powerful machine which he has used to refute some of the more ambitious conjectures in classification theory.

1.3. Amalgamation classes.

We shall review the main features of Fraïssé's theory of amalgamation classes. Much of this Memoir will be couched in terms of this theory.

DEFINITION 1.1. 1. A class $\mathcal{A}$ of finite structures is an amalgamation class if it has the following properties:
  (a) closure under isomorphism: if $A \simeq B \in \mathcal{A}$, then $B \in \mathcal{A}$;
  (b) downward closure: if $A \leq B \in \mathcal{A}$ ($\leq$ signifies "substructure"), then $A \in \mathcal{A}$;
  (c) joint embedding: if $A_1, A_2 \in \mathcal{A}$ then some structure $A \in \mathcal{A}$ contains substructures $A'_i \simeq A_i$ ($i = 1, 2$);
  (d) amalgamation: if $A_0, A_1, A_2 \in \mathcal{A}$ and $f_i : A_0 \hookrightarrow A_i$ ($i = 1, 2$) are embeddings, then A_1, A_2 can be amalgamated over A_0 in $\mathcal{A}$; that is, there is some $A \in \mathcal{A}$ and there are embeddings $g_i : A_i \hookrightarrow A$ with $g_2 f_2 = g_1 f_1$; diagrammatically:

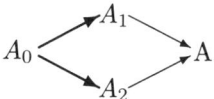

2. If Γ is a structure, then Sub (Γ) is the class of all finite structures A such that A embeds isomorphically in Γ.

PROPOSITION 1. *If Γ is a homogeneous structure, then* Sub Γ *is an amalgamation class. Conversely, if $\mathcal{A}$ is an amalgamation class of finite structures, then there is a countable homogeneous structure $\Gamma(\mathcal{A})$ with* Sub $\Gamma(\mathcal{A}) = \mathcal{A}$, *and $\Gamma(\mathcal{A})$ is unique up to isomorphism.*

To verify that Sub Γ has the amalgamation property when Γ is homogeneous, one argues as follows. Given an amalgamation problem $f_i : A_0 \hookrightarrow A_i$ ($i = 1, 2$) with $A_1, A_2 \in$ Sub Γ, we may take A_1, A_2 to be substructures of Γ, and we may suppose that f_1 is an inclusion map. As Γ is homogeneous, we may extend the isomorphism $f_2 : A_0 \simeq f_2[A_0]$ to an automorphism α of Γ. Let $A = \alpha[A_1] \cup A_2$, and define embeddings $g_i : A_i \hookrightarrow A$ by $g_1 = \alpha \upharpoonright A_1$, g_2 is the inclusion map. The triple (A, g_1, g_2) is the desired solution to the amalgamation problem.

1.3. AMALGAMATION CLASSES.

To construct a homogeneous structure associated with a specific amalgamation class, it is useful to introduce the following terminology. A structure Γ is $\mathcal{A}$-*generic* if it satisfies the following two conditions:

1. $\operatorname{Sub}\Gamma \subseteq \mathcal{A}$;
2. For $A \leq B \in \mathcal{A}$, and any embedding $f_0 : A \hookrightarrow \Gamma$, there is an extension of f_0 to an embedding $f : B \hookrightarrow \Gamma$.

It is easy to see that if $\mathcal{A}$ is an amalgamation class then there is a countable $\mathcal{A}$-generic structure Γ, and that for any $\mathcal{A}$-generic structure Γ we have $\operatorname{Sub}\Gamma = \mathcal{A}$. Our remaining claims are that for countable Γ with $\operatorname{Sub}\Gamma = \mathcal{A}$, $\mathcal{A}$-genericity is equivalent to homogeneity, and that the countable $\mathcal{A}$-generic structure is unique up to isomorphism. It is immediate that the genericity follows from homogeneity, and everything else follows from one fact:

LEMMA 1.2. *Let $\mathcal{A}$ be an amalgamation class, and suppose that Γ, Δ are countable $\mathcal{A}$-generic structures. Any isomorphism between finite structures $A \leq \Gamma$, $B \leq \Delta$ extends to an isomorphism of Γ with Δ.*

The proof is a so-called "back-and-forth" argument, modeled directly on the proof of the corresponding statement for dense linear orders. The lemma easily implies that the countable $\mathcal{A}$-generic structure is unique, and with $\Gamma = \Delta$ it says that genericity implies homogeneity.

The foregoing is adequate for the strictly utilitarian purposes of the present Memoir. What follows is intended to fill in the background more completely.

As an application of this theory, we give the proof that there are $2^{\aleph_0}$ homogeneous directed graphs. There is a natural notion of *free amalgamation* in the category of directed graphs: we may suppose the embeddings $f_i : A_0 \hookrightarrow A_i$ are inclusions and that the vertex and edge sets involved satisfy: $V(A_1) \cap V(A_2) = V(A_0)$, $E(A_1) \cap E(A_2) = E(A_0)$. Then let $V(A) = V(A_1) \cup V(A_2)$, $E(A) = E(A_1) \cup E(A_2)$. An amalgamation class which is closed under the operation of free amalgamation will be called a free amalgamation class. To any set $\mathcal{T}$ of finite tournaments we can associate a free amalgamation class $\mathcal{A}(\mathcal{T}) =:$

$\{A : \text{Every tournament embedding in } A \text{ embeds in some element of } \mathcal{T}\},$

and conversely every free amalgamation class has this form, with one possible exception: the class of directed graphs with only one vertex must be considered to be a free amalgamation class, unless the empty directed graph is admitted as a structure.

If the set $\mathcal{T}$ is an *antichain* in the sense that no tournament in $\mathcal{T}$ embeds in any other tournament in $\mathcal{T}$, then the set $\mathcal{T}$ can be recovered from $\mathcal{A}(\mathcal{T})$ as the set of maximal tournaments lying in $\mathcal{A}(\mathcal{T})$. In particular, if $\mathcal{T}$ is an infinite antichain, then the free amalgamation classes $\mathcal{A}(\mathcal{S})$, where $\mathcal{S}$ varies over arbitrary subsets of $\mathcal{T}$, are all distinct, and the corresponding countable homogeneous structures are mutually nonisomorphic. Thus if we can produce a single infinite antichain of finite tournaments, Fraïssé's theory will convert this into a supply of $2^{\aleph_0}$ countable homogeneous directed graphs.

Everything that has been said so far can be formulated abstractly whenever a suitable notion of free amalgamation is available. The class of tournaments must be replaced by the class of *amalgamation-indecomposable* structures, which are the structures A with the property that A can never embed in the free amalgam of two structures $A_1, A_2 \in \mathcal{A}$ over a common substructure A_0, unless A embeds in one of the factors A_i ($i = 1$ or 2). The concrete difficulty in all cases is to produce an antichain of amalgamation-indecomposable structures, which in algebraic situations can be quite challenging.

To produce an infinite antichain of tournaments is in itself not entirely trivial. Henson's construction is as follows. Let $L(n)$ be a linear tournament of order n, that is $L(n)$ is isomorphic to the set $\{1, \ldots, n\}$ with the edge relation: $i \longrightarrow j$ iff $i < j$. Let T_n be $L(n)$ with edges between successive vertices reversed, and with the edge from 1 to n also reversed. We claim that $\{T_n : n \geq 6\}$ is an antichain. This follows easily by examining the patterns of oriented (i.e., nonlinear) triangles C_3 embedding in T_n. Associate with T_n the symmetric graph G_n on the same vertex set in which two vertices are linked by an edge iff: in T_n, the oriented edge between them belongs to at least two distinct copies of C_3. Then by inspection G_n contains a unique cycle, of length $n-2$, for $n \geq 6$. Any embedding from T_m into T_n would induce an embedding of G_m into G_n, so we indeed have an antichain.

As has been indicated above, it emerged from [**61**] and [**57**] that Fraïssé's theory of amalgamation classes can also be used to derive limitations on the structure of homogeneous relational structures, though with considerably more effort. This point of view also makes the issues of effectivity connected with such classification problems more transparent. Although there is in general no notion of the amalgamation class *generated by* a set of finite structures, except in the case that we arbitrarily limit ourselves to a particular construction like free amalgamation, there is a useful notion of a quite similar flavor.

DEFINITION 1.3. Let $\mathcal{A}, \mathcal{B}$ be sets of finite structures.

1. The expression "& $\mathcal{A} \implies \bigvee \mathcal{B}$" means that any amalgamation class containing every $A \in \mathcal{A}$ also contains some $B \in \mathcal{B}$.
2. We say that & $\mathcal{A} \implies \bigvee \mathcal{B}$ *directly* if there is a pair of structures $A_1, A_2 \in \mathcal{A}$ and there are embeddings $f_i : A_0 \hookrightarrow A_i$ ($i = 1, 2$) of a single structure A_0 into A_1, A_2 such that any amalgam A of A_1 and A_2 over A_0 contains a structure belonging to $\mathcal{B}$.
3. A *proof* that & $\mathcal{A} \implies \bigvee \mathcal{B}$ will mean a finite string of expressions of the form & $\mathcal{A}_i \implies \bigvee \mathcal{B}_i$ in which each expression either holds directly, in the sense of clause (2), or is derivable from earlier expressions by the rule:

If & $\mathcal{A} \implies \bigvee \mathcal{B}$, and for each $B \in \mathcal{B}$, & $\mathcal{A} \cup \{B\} \implies \bigvee \mathcal{C}$, then & $\mathcal{A} \implies \bigvee \mathcal{C}$

The following "Completeness Lemma" was first stated in [**61**].

LEMMA 1.4. *Let $\mathcal{A}, \mathcal{B}$ be sets of finite structures. Then the following are equivalent:*

1. *& $\mathcal{A} \implies \bigvee \mathcal{B}$;*
2. *There is a proof that & $\mathcal{A} \implies \bigvee \mathcal{B}$.*

PROOF. Clearly ($2 \implies 1$). Now suppose there is no proof that & $\mathcal{A} \implies \bigvee \mathcal{B}$. Call a set $\mathcal{A}^*$ *consistent* if there is no proof that & $\mathcal{A}^* \implies \bigvee \mathcal{B}$. It is clear that Zorn's Lemma is applicable if a little care is taken with some set theoretical housekeeping, and that there is therefore a maximal consistent set $\mathcal{A}^*$, which we may suppose contains $\mathcal{A}$, and which necessarily does not meet $\mathcal{B}$. It is also easy to see that $\mathcal{A}^*$ is an amalgamation class: given an amalgamation problem $f_i : A_0 \hookrightarrow A_i$ ($i = 1, 2$) with $A_1, A_2 \in \mathcal{A}^*$, there are only finitely many possible solutions $B_1, \ldots, B_N$ to the problem. If none of these lie in $\mathcal{A}^*$, then by our construction there will be proofs that $\& \ \mathcal{A}^* \cup \{B_i\} \implies \mathcal{B}$ for all i, and these can easily be cobbled together into a proof that $\& \ \mathcal{A}^* \implies \mathcal{B}$. □

In particular for $\mathcal{A}, \mathcal{B}$ finite, the relation "$\& \ \mathcal{A} \implies \bigvee \mathcal{B}$" is *recursively enumerable*, that is there is an algorithm which will report correctly that this relation holds whenever it does in fact hold, and spin its wheels indefinitely otherwise.

As Lachlan pointed out, the classification problem for countable homogeneous structures in finite relational languages can be formulated in full generality quite precisely:

PROBLEM 1. *Is the relation* $\& \ \mathcal{A} \implies \bigvee \mathcal{B}$ *recursive (for finite sets of structures in finite relational languages)? In other words, is there an algorithm which decides correctly in all cases whether this relation holds?*

Of course, whether this is an adequate formulation of the classification problem is a matter of opinion. Presumably, if the answer to this Problem is negative, then the matter is settled, and if the answer turns out to be positive we should be in a better position to reformulate the problem.

We record one extremely useful observation relating to the verification that specific classes are amalgamation classes. The amalgamation property itself can present substantial difficulties, and it is worth noting that it suffices to verify the following weak form:

2-point Amalgamation: For $A_1, A_2 \in \mathcal{A}$ with a common substructure A_0, if $|A_i - A_0| = 1$ then there is $A \in \mathcal{A}$ and there are $g_i : A_i \hookrightarrow A$ with $g_1 \restriction A_0 = g_2 \restriction A_0$.

When we apply this criterion, we refer to A_0 as the common part, and the two points not in A_0 are called the "new" points. Since we shall deal exclusively with binary languages in Chapters II-V, to complete such a 2-point amalgamation diagram involves a single decision: what is the type of the pair of new points? One must bear in mind that one possibility is that the two new points will be declared equal, which of course affects the choice of the underlying set of A, as well as its structure.

From this point on we adopt the convention that all structures under consideration are countable, barring explicit mention to the contrary.

1.4. Languages, strong amalgamation, generification, and Ramsey's theorem

We shall review some elementary concepts which are useful in the discussion of homogeneous structures.

Model theorists will probably be aware of the convenience of looking at a homogeneous structure in its canonical language. For our present purposes we shall find it preferable to give a self-contained description of what we mean by a "language", recognizing that our choice varies somewhat from traditional procedures. Most readers will probably prefer to pass over these formal details.

DEFINITION 1.5.   1. A *relation* is a pair (R, n) where R is a formal symbol and n is a positive integer. R is called the relation symbol, and n is called the rank of the relation.

2. A *language* is a finite set of relations, together with a distinguished symmetric binary relation "=", and a map that carries every function

$$\imath : \{1, \dots, m\} \hookrightarrow \{1, \dots, n\}$$

into a function $\imath^*$ from relations of rank n onto relations of rank m. We require that this association be functorial: $(\jmath\imath)^* = \imath^*\jmath^*$ when defined. In particular $\operatorname{Sym}(n)$ acts on the relations of rank n.

3. If $\mathcal{L}$ is a language, an $\mathcal{L}$-structure Γ is a relational system whose relations stand in a definite 1-1 correspondence with the relations of $\mathcal{L}$, with = corresponding to equality, such that for each n such that $\mathcal{L}$ contains relations of rank n, these relations partition Γ^n, and such that for any map $\imath : \{1,\dots,m\} \hookrightarrow \{1,\dots,n\}$, the map $\imath^*$ satisfies:
The relation $\imath^*(R,n)$ holds at $a_1,\dots,a_m$ iff $\exists b_1,\dots,b_n$: the relation (R,n) holds at $b_1,\dots,b_n$, and $b_{\imath(i)} = a_i$ for all $i \leq m$.

Our main objective now is to define a very general construction, generification, which allows us to transport many homogeneous structures from their initial language to a variety of other languages. For example, the generification of a set with no structure can give rise to the Rado graph or a random tournament, by generification with varying degrees of symmetry.

DEFINITION 1.6. Let $\mathcal{A}$ be an amalgamation class. $\mathcal{A}$ is called a *strong* amalgamation class if every amalgamation problem A_i $(i=1,2)$ with $A_1, A_2 \in \mathcal{A}$ can be completed by some $g_i : A_i \hookrightarrow A$ with $A \in \mathcal{A}$ and, additionally, $g_1[A_1] \cap g_2[A_2] = g_1 \circ f_1[A_0]$.

This seems to be a fairly common occurrence, and indeed we may state the following conjecture, based on the very limited information currently available.

CONJECTURE 1.7. Let Γ be an infinite homogeneous structure for a finite binary language. If Γ is primitive then $\operatorname{Sub}\Gamma$ is a strong amalgamation class.

Let Γ be a homogeneous structure for the finite relational language $\mathcal{L}$, which we shall take to be the canonical language of Γ. Here $\mathcal{L}$ should be thought of concretely as a finite set of relation symbols, with each relation symbol carrying a specific rank (unary, binary, ...). Let $\mathcal{L}'$ be another language, and suppose $h : \mathcal{L}' \twoheadrightarrow \mathcal{L}$ is a rank-preserving surjection preserving all the functorial structure, with $h^{-1}(=)$ just $\{=\}$. We wish to define the *generification* of Γ along h; this is a homogeneous $\mathcal{L}'$-structure which intuitively should be thought of as the structure obtained from Γ by randomly replacing the relations that hold in Γ by any of their preimages. To formalize this idea it is convenient to pass to the amalgamation class $\mathcal{A} = \operatorname{Sub}\Gamma$, which is a class of finite $\mathcal{L}$-structures. If A is an $\mathcal{L}'$-structure, let h_*A be the $\mathcal{L}$-structure obtained by replacing each relation holding in A by its image under h. Let $h^*\mathcal{A}$ be $\{A \text{ an } \mathcal{L}'\text{-structure} : h_*A \in \mathcal{A}\}$.

PROPOSITION 2. Let $h : \mathcal{L}' \twoheadrightarrow \mathcal{L}$ as above, and suppose that $\mathcal{A}$ is a strong amalgamation class of $\mathcal{L}$-structures. Then $h^*\mathcal{A}$ is a strong amalgamation class of $\mathcal{L}'$-structures.

PROOF. Given an amalgamation problem $f_i : A_0 \hookrightarrow A_i$ $(i=1,2)$ in $h^*\mathcal{A}$, we obtain an amalgamation problem $f_i : h_*A_0 \hookrightarrow h_*A_i$ in $\mathcal{A}$, with solution $g_i : h_*A_i \hookrightarrow A$, where $A \in \mathcal{A}$ satisfies: $g_1 \circ f_1[A_1] \cap g_2 \circ f_2[A_2] = g_1 \circ f_1[A_0]$ (as sets). It is easy to see that there is then a $\mathcal{L}'$-structure A^* with the same elements as A, such that

1.4. STRONG AMALGAMATION, GENERIFICATION, RAMSEY

$g_i[A_i]$ is a substructure of A^* for $i = 1, 2$, and $h_* A^* = A$. Then A^* solves the original amalgamation problem. □

When the foregoing proposition applies, we can use it to speak of the *generification* of Γ along h, by which we mean the homogeneous $\mathcal{L}'$-structure $h^*\Gamma$ associated with $h^*\operatorname{Sub}\Gamma$, when $\operatorname{Sub}\Gamma$ is a strong amalgamation class. Then $h_* h^* \Gamma \simeq \Gamma$.

For example, if $\mathcal{L}$ is the binary language containing only equality and inequality (and a degenerate unary relation), and we cover $\mathcal{L}$ by letting two binary relations cover the relation $\neq$, then depending on the way we extend the action of $\operatorname{Sym}(2)$ (symmetrically or asymmetrically) the generification will be either the Rado graph or the random tournament.

We shall now discuss Ramsey's theorem briefly. We state the binary case of this principle in both the finite and infinite versions.

THEOREM 1.8. *Let c, n be fixed and finite. There is an N such that any edge coloring of the complete graph K_N by c colors is monochromatic on some induced subgraph of order n.*

THEOREM 1.9. *Any edge coloring of an infinite complete graph by finitely many colors is monochromatic on some infinite induced subgraph.*

We shall need to use the finite version extensively in connection with an amalgamation trick due to Lachlan. At the same time, the infinite version has one strong and suggestive consequence.

DEFINITION 1.10. Let Γ be a homogeneous structure for a finite binary language, R a substructure of Γ. Then R is a *ramsey substructure* if there is a linear ordering of R such that the pairs $\{(a, b) : a, b \in R, a < b\}$ all lie in one orbit under $\operatorname{Aut}\Gamma$. (Model theorists would call R an indiscernible sequence.)

COROLLARY 1. *Let Γ be an infinite structure in a finite binary language. Then Γ contains an infinite ramsey substructure.*

PROOF. Let $K = \Gamma$ as a set, and think of K as an infinite complete graph. Equip K with a linear ordering $<$ in any fashion, and color the edges of K by assigning to the edge ab the orbit of (a, b) if $a < b$. This edge coloring uses finitely many colors, and any infinite monochromatic induced subgraph is a ramsey structure, with the given linear ordering serving to witness this fact. □

Remarkably, this style of argument is one of the main tools in the explicit classification of infinite homogeneous binary structures. All of this generalizes at once to finite relational languages which are not assumed binary, but it has not as yet proved terribly useful in this more general context.

Finally, we introduce some notation common in model theory which will prove very convenient throughout the analysis in Chapters 2-8.

Let Γ be a homogeneous binary structure, $a \in \Gamma$, $X \subseteq \Gamma$. The *type* of a over X is the isomorphism type of the structure $X \cup \{a\}$ over X. In the present context this is determined by the isomorphism types of all the pairs (a, x) with $x \in X$, or equivalently by the orbits of these pairs under $\operatorname{Aut}\Gamma$. The type of a over X is denoted $\operatorname{tp}(a/X)$, and when $X = \{b\}$ it is denoted by either of the following: $\operatorname{tp}(a/b)$ or (more commonly) $\operatorname{tp}(ab)$. Thus it is not too great an abuse of notation to write: $\operatorname{tp}(a/X) = \{\operatorname{tp}(ax) : x \in X\}$, and this is certainly a good way to think about it.

In a similar vein, a *2-type* is one of the orbits of $\operatorname{Aut}\Gamma$ on Γ^2, and if p is a 2-type and $a \in \Gamma$, then $a^p = \{b : (a,b) \in p\}$, $^p a = \{b : (b,a) \in p\}$.

Later we shall modify the notion of 1-type in the context of 2-tournaments and 2-directed graphs. When a structure is partitioned into two distinguished subsets we shall mainly deal with 1-types realized by points of the first set over points of the second set, for purely technical reasons, and we shall therefore restrict the terminology accordingly.

1.5. Classification theorems.

We shall now give a fairly compact description of the classification of the homogeneous n-tournaments and homogeneous directed graphs. More detailed statements will be found at the beginning of Chapters 2 and 5.

DEFINITION 1.11. An *n-tournament* is a structure $\mathbb{T}$ consisting of a set of vertices partitioned into n distinguished subsets $V_1, \ldots, V_n$ of $V(\mathbb{T})$, some of which may be empty, and with a finite number of types of "edges" (binary relations) associated to each pair $\{i,j\}$ with $1 \leq i \leq j \leq n$; when $i = j$ there is a unique asymmetric edge relation E_i, and $T_i = (V_i, E_i)$ is a tournament, while for $i < j$ there are finitely many edge relations partitioning $V_i \times V_j$ (it makes no difference whether we view these relations as asymmetric — disjoint from $V_j \times V_i$ — or symmetric — extended canonically to $V_j \times V_i$).

We write informally $\mathbb{T} = (T_1, \ldots, T_n)$, a substantial abuse of notation.

Observe that a 1-tournament is indeed a tournament.

NOTATION 1. *Given an n-tournament $\mathbb{T} = (T_1, \ldots, T_n)$ and a subset I of the index set $\{1, \ldots, n\}$, the restriction $\mathbb{T} \restriction I$ is the $|I|$-tournament induced on $\bigcup_{i \in I} T_i$ by $\mathbb{T}$. For $m \leq n$, the m-restrictions of $\mathbb{T}$ are the I-restrictions with $|I| = m$.*

The critical result states that a homogeneous n-tournament is almost completely determined by its 2-restrictions. To state this precisely we require the following notion.

DEFINITION 1.12. Given an n-tournament $\mathbb{T}$, a *transversal* of $\mathbb{T}$ is an n-subtournament $\mathbb{S}$ with each component S_i containing at most one vertex.

THEOREM 1.13. *A homogeneous n-tournament is determined up to isomorphism by the following data: the isomorphism types of its transversals, and the isomorphism types of its 2-restrictions.*

The two types of data involved are not independent. We shall supplement this result by an explicit description of all homogeneous 2-tournaments at the beginning of Chapter II, after which it will be possible to state our classification theorem more explicitly. We note that this more precise version of the theorem implies, as one would expect, that the relation $\& \mathcal{A} \implies \bigvee \mathcal{B}$ examined in §3 is effectively decidable in the context of n-tournaments. The classification also implies that there are only countably many homogeneous n-tournaments, since the foregoing theorem reduces this question to the case $n = 2$, for which we shall give a complete list. Our second result involves the classification of an uncountable family of homogeneous structures.

THEOREM 1.14. *A homogeneous directed graph H is either associated with a free amalgamation class, in which case it is completely determined by the set of tournaments embedding in H, or is one of a countable number of known homogeneous directed graphs which fall naturally into finitely many parametrized families.*

A preliminary catalog of the homogeneous directed graphs known in 1983 was given in [16], including one that emerged from a rather superficial but systematic analysis (influenced by [57]) connected with the preparation of that catalog. In working out the full classification, two new examples were discovered, an imprimitive one described in [17], and a primitive one which has not previously been described. We give a full description of all the examples at the beginning of Chapter 5. The classification implies that the relation $\& \ \mathcal{A} \implies \bigvee \mathcal{B}$ described in §3 is effectively decidable in the context of directed graphs. As we have explained in §3, it has been known for the past two decades that there are $2^{\aleph_0}$ homogeneous directed graphs associated with free amalgamation classes.

1.6. Open problems.

We feel that the central problem in the model theory of homogeneous structures is the one discussed by Lachlan in [59] and presented in §3 as Problem 1:

> Is the relation $\& \ \mathcal{A} \implies \bigvee \mathcal{B}$ recursive (for finite sets of structures in finite relational languages)? In other words, is there an algorithm which decides correctly in all cases whether this relation holds?

The following closely related problem, also discussed in [59], is considerably more concrete. If $\mathcal{N}$ is a class of finite structures (of a given type), let $\mathcal{A}(\neg\mathcal{N})$ be the class of all finite structures A (of the same type) such that no structure in $\mathcal{N}$ embeds isomorphically in A. In this context we call $\mathcal{N}$ a set of *negative constraints*, and we say that $\mathcal{A}(\neg\mathcal{N})$ is the class of structures *satisfying* the given constraints. Any downward-closed class can be presented as a class defined by negative constraints. The important notion is the following: a class $\mathcal{A}$ is *finitely constrained* if $\mathcal{A} = \mathcal{A}(\neg\mathcal{N})$ for some finite set $\mathcal{N}$.

PROBLEM 2. *Given a finite relational language $\mathcal{L}$, is every amalgamation class of $\mathcal{L}$-structures the intersection of a decreasing family of finitely constrained amalgamation classes?*

It is noted in [59] that for binary languages, a positive answer to the second problem yields a positive answer to the first. It is not clear how to obtain such a reduction for more general languages.

We have already noted that the first problem is solved for n-tournaments and for directed graphs by our two classification theorems, and it is almost inevitable that any respectable classification theorem would yield a positive solution to the corresponding case of the first problem. The second problem also has a positive solution in these two settings. The free amalgamation classes are always presentable as decreasing intersections of finitely constrained free amalgamation classes, and in the two cases considered here, all of the homogeneous structures correspond to amalgamation classes which are either finitely constrained, or can be viewed as free amalgamation classes.

Viewed in the light of these general problems, the information obtained in the present Memoir has to be viewed as anecdotal; it has not brought us out of the dark ages, but it seems fair to say that it brings us *into* the dark ages. For various reasons, more information along the same lines would be desirable. Some mildly new types of examples have been uncovered by the analysis so far, and it is possible that some really new phenomena would emerge if the classification were pushed a little farther. Of course, if the methods we have been using were to break down, this would be extremely interesting. It is also possible that the essential features of the analysis have yet to emerge. The main problems involved in a direct continuation of this line of research are the following.

PROBLEM 3. *Classify the homogeneous n-graphs and n-digraphs.*

This problem seems technically quite difficult. For each n there are only finitely many homogeneous n-tournaments. This fact has no analog for n-graphs, even if the isomorphism types of the individual components are fixed. This is because there are $2^{\aleph_0}$ free amalgamation classes of 2-graphs once one has at least 3 cross types, since there is an antichain of amalgamation-indecomposable finite 2-graphs in this case.

It would also be interesting to continue the work on homogeneity of directed graphs by an analysis of two other concrete cases: complete graphs with a 3-edge coloring, and tournaments with a 2-edge coloring (so that there are four possible relations between two distinct points). Both of these cases should be more delicate than the case of directed graphs, though much of our overall approach is readily transferred to these contexts. It would be prudent, though not absolutely necessary, to work out the classification of the homogeneous n-graphs before attacking the homogeneous edge-colored graphs. It should be noted that the Fraïssé construction produces $2^{\aleph_0}$ examples in each of these cases. It follows that there are $2^{\aleph_0}$ examples in any case more complex than that of graphs.

There is one very narrow problem which would seem to call for a separate investigation.

PROBLEM 4. *Let $\mathcal{L}$ be a finite binary language. Let $\mathcal{N}$ be a class of $\mathcal{L}$-structures with three vertices, and let $\mathcal{A}(\neg \mathcal{N})$ be the associated class of $\mathcal{L}$-structures, negatively constrained by $\mathcal{N}$. Give a good criterion for $\mathcal{A}(\neg \mathcal{N})$ to be an amalgamation class.*

See the Appendix for some examples.

On the one hand, this problem is effectively decidable in a routine manner, and one can simply test directly whether $\mathcal{A}(\neg \mathcal{N})$ is an amalgamation class. At the same time one could want to understand the situation somewhat better, since in our present state of knowledge, whenever one attempts an explicit classification project, the inductive argument requires rather precise knowledge of the simplest examples; it does not seem to be enough to know that all such examples are readily identified in principle. In fact all of the special examples known to date are connected in one way or another with homogeneous structures determined by negative constraints of order 3, or are finite. Of course homogeneous structures determined by negative constraints of order 3 can correspond to certain free amalgamation classes, or be imprimitive, or finite; there are a number of other examples connected with partial or total orders, as well as other examples which appear unconnected with any familiar mathematical construction; for example, in a language with three symmetric nontrivial binary relations (representing 2-types) there is essentially one primitive amalgamation class given by negative constraints of order three which is not a free amalgamation class (unique up to permutation of the relation symbols).

1.6. OPEN PROBLEMS.

For a binary homogeneous structure determined by negative constraints of bounded order N to be finite, it is necessary and sufficient that all the "monochromatic" structures of order N be excluded: here a monochromatic structure is one that carries a linear order such that any pair of points a, b with $a < b$ satisfy the same relations.

In all of this work it is quite troublesome that many extremely elementary but arduous lemmas dealing with individual relations of the form $\& \, \mathcal{A} \implies \bigvee \mathcal{B}$ must be dealt with. In practice one knows very quickly what these lemmas must say, and even before the lemmas are checked one can see that they have the following two properties:

1. The specific lemmas are jointly equivalent to the original classification conjecture;
2. Each such lemma, if true, can be verified by a mechanical procedure.

This suggests that significant portion of the work involved in this sort of investigation could be automated. Unfortunately it is not clear how this should be done. A very direct approach would suffer exponential blowup, but in practice a little thought and some experimentation produces a relatively short proof.

PROBLEM 5. *Is there a practical algorithm which recognizes most true implications* $\& \, \mathcal{A} \implies \bigvee \mathcal{B}$ *in small binary languages, and delivers the data needed to construct short proofs of such facts?*

A number of good tests for such an algorithm are provided by the lemmas of Chapter 6. The general flavor of the problem seems like computer chess.

These are the main problems from our present point of view. There are of course innumerable other problems connected with the matters touched on lightly at the end of §2. Some of the combinatorial problems arising in the subject have been described in [19], including some problems in finite graphs arising from an attempt to understand the known homogeneous graphs somewhat better. In [19] we also mentioned one algorithmic problem connected with the classification of homogeneous structures which is still open in the context of homogeneous directed graphs in spite of the existence of a full classification. That problem runs as follows.

Given two finite sets $\mathcal{P}, \mathcal{N}$ of finite directed graphs, we may treat them as positive and negative constraints respectively, and ask for (countable) homogeneous structures embedding all structures in $\mathcal{P}$ and no structures in $\mathcal{N}$. Our classification then gives us a clear picture of the possibilities. On the other hand, as soon as we ask whether this collection of (isomorphism types of) homogeneous structures is countable or uncountable, it turns out that there is no obvious algorithmic solution. The problem reduces quickly to a problem about tournaments: given a finite set $\mathcal{T}$ of finite tournaments, to determine whether there is an infinite set $\mathcal{T}^\perp$ of finite tournaments, such that $\mathcal{T} \cup \mathcal{T}^\perp$ is an antichain. My former student Brenda Latka found an explicit solution in the case that $|\mathcal{T}| = 1$. When there are no infinite antichains, this can be proved by a reduction to Kruskal's theorem. Furthermore, in every case in which there is an infinite antichain, one of two antichains given by Latka works (after deleting some finite set of elements). In 1996 Latka and the present author proved an ineffective finiteness theorem relating to this problem which has some effective consequences. In particular with some modest additional work it yields an effective solution to the problem when $|\mathcal{T}| = 2$, which relies heavily on the detailed solution for $|\mathcal{T}| = 1$.

We recall one other issue raised in [19]. Let $\mathcal{A}$ be a finitely constrained amalgamation class. For each finite n, let $\mathcal{A}_n$ be the set of all structures whose underlying set is the fixed set $\{1, \ldots, n\}$, and which lie in $\mathcal{A}$. It would be interesting to know whether the sequence $\mathcal{A}_n$ satisfies a 0-1 law for asymptotic probabilities of first order sentences in the style of [30, 38, 52, 53], particularly when $\mathcal{A}$ is a free amalgamation class, and if so, what the limit theory looks like, where we use a uniform distribution on the set $\mathcal{A}_n$. The background to this is discussed more fully in [19]. The striking work of [28, 52, 53] does not appear to go over to the directed case, even in the simple case in which $\mathcal{A}$ is characterized by omission of one of the two tournaments of order 3.

It would be interesting to settle the following special cases.

CONJECTURE 1.15. A random large finite directed graph containing no copy of $L(3)$ (the linear tournament on three vertices) is tripartite with probability near 1.

CONJECTURE 1.16. If Γ is a random large finite directed graph omitting the nonlinear tournament C_3 on three vertices, then with probability close to 1 there is an ordering of the vertices such that for all $a, b \in V(\Gamma)$, if $a \longrightarrow b$ then $a < b$.

CHAPTER 2

Homogeneous 2-tournaments

2.1. A catalog.

We begin with an explicit description of all homogeneous n-tournaments. We shall prove in this chapter and the next that our catalog is complete. The present chapter treats the case of 2-tournaments, modulo a lemma on 3-tournaments whose proof will be deferred to the chapter following.

We defined the class of n-tournaments precisely in section I.5. Recall the informal notation $\mathbb{T} = (T_1, \ldots, T_n)$ introduced at that time, the notion of an I-restriction, for $I \subseteq \{1, \ldots, n\}$, and the notion of an m-restriction.

One way to get an n-tournament is to take an ordinary tournament with a vertex coloring, and use the orientation of the edges to divide the edges between vertices of distinct colors into two groups. These are the n-tournaments used by Lachlan in [57]; they have only two 2-types between any two distinct components. As it happens, the general features of the classification of n-tournaments first emerge when one allows at least 3 cross types (types of edges between distinct components).

Our main result will show that a homogeneous n-tournament is almost determined by the isomorphism types of its 2-restrictions, so we begin by giving explicit descriptions of the homogeneous 1-tournaments and 2-tournaments before describing the general case.

We shall make use of the following notation. For T a tournament, and $a \in T$, let $a' = \{b \in T : a \longrightarrow b\}$ and $'a = \{b \in T : a \longleftarrow b\}$; or equally often, as the context determines: let $a', 'a$ denote the tournaments induced on these two sets. One reason this is useful is that in a homogeneous tournament, the tournaments $a', 'a$ are again homogeneous, and their isomorphism types are independent of the choice of a.

We shall also say that a *dominates* b when $a \longrightarrow b$ holds.

CATALOG 1. The homogeneous tournaments [57].

1. I_1 The 1-point tournament.
2. C_3 The oriented 3-cycle (the unique nonlinear tournament on three vertices).
3. $\mathbb{Q}$ The rational order, viewed as a tournament via:
   $$q \longrightarrow r \text{ iff } q < r.$$
4. $S(2)$ The dense *local order*. The vertices are a countable dense set of points on the unit circle containing no antipodal pairs, and edges are drawn counterclockwise up to half-way around.
5. T^∞ The generic tournament associated with the amalgamation class of all finite tournaments.

Comment

A *local order* may be defined as a tournament T in which for all vertices a, the associated tournaments $a',\,'a$ are linear. S(2) is the unique countable dense local order. We shall encounter the homogeneous directed graph S(3) below, at which point the reason for this choice of notation will become clearer.

It is rather easy to classify the homogeneous local orders, getting the first four tournaments on our list. The whole difficulty in [**57**] was to characterize T^∞ as the unique homogeneous tournament which is not a local order; as noted earlier, the streamlined account of this in [**18**] illustrates the use of the techniques from [**57**] in the form they take in the present Memoir.

As far as the verification that these tournaments are indeed homogeneous is concerned, only the fourth presents any difficulty, and this can easily be handled by verifying that a dense local order is generic for the class of all finite local orders, in the sense of §1.3. In this connection it should be noted that $\mathcal{A}$-genericity of a structure Γ is equivalent to the ostensibly weaker condition:

For every pair of structures $A \leq B \in \mathcal{A}$, with $|B - A| = 1$, and any embedding $f : A \hookrightarrow \Gamma$, f extends to an embedding of B into Γ.

Iterated applications of this condition yield full genericity.

The main result of [**57**] is that every homogeneous tournament is on the foregoing list. It should also be noted that the homogeneous tournaments which correspond to strong amalgamation classes in the sense of §1.3 are exactly the three infinite tournaments, and that all five examples are primitive.

We shall now describe the homogeneous 2-tournaments. Here instead of giving a catalog it is more natural to think in terms of constructions of n-tournaments from tournaments.

DEFINITION 1 (Diagonal n-tournament). Let T be a tournament, $n \geq 2$. The *diagonal n-tournament* $\mathbb{T} = \Delta_n(T)$ is a tournament with n components T_i isomorphic to T, related by a coherent family of isomorphisms ι_{ij} between the siblings, that is, ι_{ii} is the identity, and $\iota_{jk} \circ \iota_{ij} = \iota_{ik}$; $\mathbb{T}$ is equipped with the following binary relations:

$$a \in T_i, b \in T_j, \quad \iota_{ij}(a) = b;$$
$$a \in T_i, b \in T_j, \quad \iota_{ij}(a) \longrightarrow b;$$
$$a \in T_i, b \in T_j, \quad \iota_{ij}(a) \longleftarrow b;$$

(as well as the edge relation in each component). Then $\mathbb{T}$ is a homogeneous n-tournament.

Together with the diagonal n-tournament as constructed here, any variant in which the orientations of the edges are reversed in some components is also considered to be a diagonal n-tournament.

DEFINITION 2. [Shuffling] Let T be one of the dense homogeneous local orders, $\mathbb{Q}$ or S(2), and let $n \geq 2$. Partition T into n dense pieces $T_1, \ldots, T_n$. The n-tournament $\mathbb{T} = (T_1, \ldots, T_n)$, with the edge relation inherited from T, is called a *shuffled n-tournament of type T*. Any variant in which the edge orientations are reversed in some components is also considered to be a shuffled n-tournament.

We immediately generalize this as follows. The canonical language for $\mathbb{T}$ contains two nontrivial binary relations $\underset{ij}{\longrightarrow}$, $\underset{ij}{\longleftarrow}$, between each pair of components T_i, T_j. As described in §1.4, we can form the *generification* of $\mathbb{T}$ with respect to any covering of the relations on $\mathbb{T}$; we choose to leave the relations in each component intact, but to replace the relations between distinct components T_i, T_j; one way to describe this would be to associate to each pair i, j with $i < j$ two sets of relations L_{ij}, R_{ij} standing for "left" and "right", letting L_{ij} cover $\longrightarrow_{ij}$, and R_{ij} cover $\longleftarrow_{ij}$ (and then complete the language to allow the action of Sym(2)).

Our final construction involves a more delicate sort of generification used to construct 2-tournaments. The specification of a language for 2-tournaments depends primarily on a parameter k, the number of relations holding in $T_1 \times T_2$, which we refer to as the *cross types*; we also have the two unary relations picking out the components, and the asymmetric edge relation within each component. The action of Sym(2) on the binary relations and the projections from binary to unary relations involve no significant choices.

DEFINITION 3. [$\mathcal{P}$-generic 2-tournaments] Let P be a set of $k \geq 1$ cross types (binary relation symbols). Let $\mathcal{P} = (P^0, P^+, P^-)$ be a partition of P into three sets, and let T_1, T_2 be two homogeneous tournaments. We define $\mathcal{A}(T_1, T_2; \mathcal{P})$ as the class of finite 2-tournaments $\mathbb{A} = (A_1, A_2)$ such that:

1. The cross types of $\mathbb{A}$ come from P;
2. A_1, A_2 embed in T_1, T_2 respectively;
3. There are no substructures of $\mathbb{A}$ on three vertices of the form:

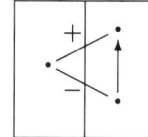 or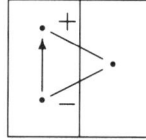

where $+, -$ denote arbitrary cross types in P^+, P^- respectively. That is, over any point a in one component, all of the points related positively to a dominate all of the points related negatively to a.

If $\mathcal{A}(T_1, T_2; \mathcal{P})$ is an amalgamation class, the associated homogeneous 2-tournament is said to be *$\mathcal{P}$-generic*. Any variant in which the orientations of the edges are reversed in one of the components is also considered $\mathcal{P}$-generic.

In fact we shall only apply this construction in two special cases. If $P^0 = P$ then this construction works with any two infinite components, and just gives the generification along P of the unique 2-tournament with the specified components in which $k = 1$; as noted earlier, the infinite tournaments correspond to strong amalgamation classes. In this case we shall call the resulting 2-tournament $\Gamma_k(T_1, T_2)$. Of course this construction also works with one or both components finite if $k = 1$, and we shall use the same notation in this degenerate case.

The second variation on this construction will apply when $T_1 \simeq T_2 \simeq T^\infty$, and P^0 is nonempty. In this case we shall denote the corresponding 2-tournament by $\Gamma(\mathcal{P})$. It is certainly necessary to verify that $\mathcal{A}(\mathcal{P})$ will have the amalgamation property in this case, but as noted at the end of §1.3, we may restrict our attention to amalgamations involving only two points lying outside the common part $\mathbb{A}_0$. We distinguish two cases:

Case 1. Suppose the two new points $c_1 \in \mathbb{A}_1, c_2 \in \mathbb{A}_2$ lie in the same component. Since that component is T^∞, the only possible difficulty that could arise would be the following: there are points a, b in the other component so that the cross-type of c_1 over a is in P^+, while its cross-type over b is in P^-, and the situation is reversed for c_2. Examining c_1, a, b in $\mathbb{A}_1$ we discover that $a \longrightarrow b$. Passing to $\mathbb{A}_2$ we discover the opposite. So this difficulty does not arise.

Case 2. Suppose the two new points lie in distinct components. Then any type from P^0 can be used to complete the diagram.

THEOREM 2.1. *A homogeneous 2-tournament is either diagonal ($k = 3$), shuffled (with both components isomorphic to $\mathbb{Q}$, or both isomorphic to $S(2)$), or $\mathcal{P}$-generic of one of the types described above: either $\Gamma_k(T_1, T_2)$ with T_1, T_2 both infinite, or $k = 1$, or else $\Gamma(\mathcal{P})$ with P^0 nonempty.*

We now recall the statement of the main theorem on homogeneous n-tournaments.

THEOREM 2.2. *A homogeneous n-tournament is determined up to isomorphism by the following data: the isomorphism types of its transversals, and the isomorphism types of its 2-restrictions.*

We now discuss the question of effectivity in some detail. With an index set I of size n fixed, suppose that we are given the following data:

1. For $i, j \in I$, the isomorphism type of a 2-tournament; in particular we have a specified number k_{ij} of cross types;
2. A set of allowed transversals (isomorphism types of n-tournaments involving the specified cross types, in which every component has at most one element).

The class $\mathcal{A}$ of finite n-tournaments *compatible* with the given data is defined in the obvious way as those whose 2-restrictions embed in the specified 2-restrictions, and whose transversals are allowed. We need to see that it is possible to determine whether a given set of data actually determines an amalgamation class. The point is that our data determine a finitely constrained class, because the possible 2-restrictions all happen to be finitely constrained (by inspection). Furthermore in the binary setting it is always possible to check effectively whether a finitely constrained class is an amalgamation class, as was noted in [59]. That is, as explained in §1.3, we need only see whether every 2-point amalgamation problem from $\mathcal{A}$ can be solved in $\mathcal{A}$, and these problems have the form A_0, A_1, A_2, with $A_1, A_2 \in \mathcal{A}$ and $A_i = A_0 \cup \{a_i\}$ for $i = 1, 2$. It is therefore sufficient to find a bound N such that if the given problem has no solution in $\mathcal{A}$, then for some $B_0 \subseteq A_0$ with $|B_0| \leq N$, the problem B_0, B_1, B_2 with $B_i = B_0 \cup \{a_i\}$ ($i = 1, 2$) also has no solution; and such a bound is trivial in the finitely constrained binary case.

This rather abstract analysis can be supplemented by some more practical considerations which reduce the amount of effort involved in checking the feasibility of a set of data considerably.

Begin with a homogeneous n-tournament $\mathbb{T}$, with components indexed by $I = \{1, \ldots, n\}$. We define two equivalence relations D, S on I as follows: $D(i,j)$ holds if (T_i, T_j) is diagonal; $S(i,j)$ holds if (T_i, T_j) is either shuffled or diagonal. To see that these are in fact equivalence relations, observe that (T_i, T_j) is diagonal just in case there is an (Aut $\mathbb{T}$)-invariant isomorphism or antiisomorphism between T_1 and T_2, and $S(i,j)$ holds just in case there is an (Aut $\mathbb{T}$)-invariant isomorphism or antiisomorphism of T_i with a dense subset of the dedekind completion of T_j.

We can suppose that the relation D is trivial, since there is an essentially unique way, up to linguistic variation, to recover $\mathbb{T}$ from its restriction to a set of representatives for the D-classes. $\mathbb{T}$ may then be described as follows:

1. Specify the S-classes, and the isomorphism types of the shuffled 2-restrictions within each class; this imposes further constraints on the transversals consisting of three points from distinct components whose indices lie in a single S-class;
2. Specify the set $I_\infty = \{i \in I : T_i \simeq T^\infty\}$. For i, j in I_∞ there is a well-defined partition $\mathcal{P}_{ij}$ of the i, j-cross types P_{ij} into three sets $P_{ij}^0, P_{ij}^+, P_{ij}^-$ with P_{ij}^0 nonempty, such that $(T_i, T_j) \simeq \Gamma(\mathcal{P}_{ij})$. There are additional constraints on triangles whose vertices lie in distinct components indexed by $i, j, k \in I_\infty$. Let $P_{ijk}(+-)$ denote the set of cross types p in P_{ik} for which there is a triangle with vertices in T_i, T_j, T_k, with the cross-type in P_{ik} equal to p, and the other two cross types in P_{ij}^+, P_{jk}^- respectively; define $P_{ijk}(-+)$ similarly. Then we require $P_{ijk}(+-) \cap P_{ijk}(-+) = \emptyset$. (So this is a condition on the set of transversals.)
3. Complete the description of the allowed transversals, bearing the constraints obtained in (1,2) in mind, and of course insisting that the allowed constraints be closed downward (it is convenient, as we have noticed, to speak of transversals with some components empty).

Such data will not automatically correspond to an amalgamation class, but we seem to have recorded all of the generally useful requirements, and it seems that some considerations along the lines of those invoked at the outset will be needed to finish the analysis in general.

Since the homogeneous tournaments are known [57], we begin our analysis with the 2-tournaments. There is a classification of 2-tournaments with $k = 2$ in [57], but we shall not use it.

In the next section we treat the more special varieties of homogeneous 2-tournaments separately.

Most of the notation we need for tournaments and 2-tournaments has now been introduced. We record our notation for a few more useful objects and constructions.

$L(n)$: a linear tournament with n vertices;

$T_1[T_2]$: the *composition* or *wreath product* of T_1, T_2: each vertex of T_1 is replaced by a copy of T_2, and edges between distinct copies of T_2 are controlled by the edges of T_1;

$[T_1, T_2]$: the disjoint union of T_1, T_2, extended to a tournament by taking $a \longrightarrow b$ whenever $a \in T_1, b \in T_2$.

2.2. Restricted homogeneous 2-tournaments.

We shall call a homogeneous 2-tournament $\mathbb{T}$ *restricted* if it has a component not isomorphic to T^∞, or if it is diagonal. In the present section we classify these 2-tournaments. In the proof of Proposition 5 we shall require the following result on 3-tournaments.

LEMMA 2.3. *Let $\mathbb{T}$ be a homogeneous 3-tournament of type $(\mathbb{Q}, \mathbb{Q}, S(2))$ with (T_1, T_2) shuffled, and with the $\{1, 3\}$- and $\{2, 3\}$-restrictions of $\mathbb{T}$ both isomorphic to $\Gamma_k(\mathbb{Q}, S(2))$. Suppose also that $\mathbb{T}$ embeds all possible transversals (that is, all transversals involving suitable cross types). Then $\mathbb{T}$ is generic relative to the specified constraints, that is it embeds every finite 3-tournament compatible with the specified constraints.*

We do not wish to pause here to prove this, as it is a rather special instance of the classification of n-tournaments in terms of 2-tournaments and transversals, and as such will be dealt with in a later section. There is some danger of circularity, so we observe that this result will not be invoked until we have already classified the homogeneous 2-tournaments with one component isomorphic to $\mathbb{Q}$, and the other component isomorphic to $\mathbb{Q}$ or $S(2)$, which are the only cases which arise in the proof of the foregoing Lemma.

LEMMA 2.4. *Let $\mathbb{T}$ be a homogeneous 2-tournament with T_1 finite. Then $\mathbb{T}$ is either diagonal or of the form $\Gamma_k(T_1, T_2)$ with $k = 1$.*

PROOF. All homogeneous tournaments are primitive, so the equivalence relation $E(a,b)$ defined on T_2 by the relation $\mathrm{tp}(a/T_1) = \mathrm{tp}(b/T_1)$ is either discrete (equality) or degenerate (vacuous). In the former case T_2 is also finite, and the claim holds by inspection, while in the latter case $k = 1$ and T is k-generic. □

PROPOSITION 3. *Let $\mathbb{T}$ be a homogeneous 2-tournament with $T_1 \simeq \mathbb{Q}$ or $S(2)$. Suppose that for some $b \in T_2$ and some cross-type p, $^p b$ is not dense in T_1. Then $T_2 \simeq T_1$ and $\mathbb{T}$ is a shuffled or diagonal 2-tournament.*

PROOF. Our assumption on b and p yields a cross type q and elements a_1, a_2 in $^q b$ with $a_1 \longrightarrow a_2$ such that the interval $(a_1, a_2) = \{a \in T_1 : a_1 \longrightarrow a \longrightarrow a_2\}$ is disjoint from $^p b$. The convex closure $C(^q b)$ of $^q b$ is then disjoint from $^p b$, and is in particular proper and linear.

We now show that $C(^q b)$ is a *cut* in T_1, by which we mean either the left or right side of an ordinary dedekind cut in T_1 if $T_1 \simeq \mathbb{Q}$, or a maximal linearly ordered subset of T_1 if $T_1 \simeq S(2)$. In the contrary case, there would be $a_1 \longrightarrow a_2 \in T_1$ with $C(^q b) \subseteq a'_1 \cap\, 'a_2$. Then for $a_3 \longrightarrow a_4$ chosen in $C(^q b)$, there would be b_1 with $C(^q b_1) \subseteq a'_3 \cap\, 'a_4 \subseteq C(^q b)$. This easily gives rise to at least three 2-types in T_2: the types of (b, b_1), (b_1, b) and a pair (b_1, b_2) with $C(^q b_1)$ disjoint from $C(^q b_2)$. So this is a contradiction.

If $C(^q b)$ coincides with a' or $'a$ for some $a \in T_1$, then we shall say the cut is *rational*. In this case we have a canonical identification of T_2 with T_1 associating each $b \in T_2$ with the corresponding $a \in T_1$. This leads at once to the identification of $\mathbb{T}$ as a diagonal 2-tournament. Now assume that the cuts $C(^q b)$ are irrational. We have an embedding of T_1 into the dedekind completion of T_2 which is either an isomorphism or an antiisomorphism, and we may assume that it is an isomorphism. Observe now that for any cross-type p, $C(^p b)$ is defined, and is either $C(^q b)$ or its complement. Thus the cross types may be divided into two classes, which we call L and R. If $T_1 \simeq T_2 \simeq \mathbb{Q}$ then we may suppose that $x^L < x^R$ for $x \in T_1$, where e.g. $x^L = \{y \in T_2 : \mathrm{tp}(x, y) \in L\}$. In this case we find that for $y \in T_2$ we have the opposite relation $^L y > {}^R y$. If on the other hand $T_1 \simeq T_2 \simeq S(2)$, then for any pair $x \longrightarrow y$ in T_1, we may suppose that $y^L \cap x^R < y^R \cap x^R$; the opposite inequality then holds in x^L.

Our claim is of course that $\mathbb{T}$ is an (L,R)-shuffled 2-tournament. Now it is easy to see that for $x \in T_1$ and $r \in R$, the set x^r is dense in x^R; and similarly for types $l \in L$. Therefore if $T \leq T_2$ is finite, then any consistent 1-type over T must be realized in T_1. But this already implies that $\mathbb{T}$ is the generic (L, R)-shuffle of T_1 and T_2, because any finite configuration (A, B) which can be embedded in the generic shuffle can be built up as the uniqe object obtainable by amalgamating a family of realizations of 1-types (in T_1) over a finite subset B^* of T_2. Here B^* would be the

2.2. RESTRICTED HOMOGENEOUS 2-TOURNAMENTS.

union of B with finitely many additional parameters related to A in such a way that the ordering on A is determined by the types of the elements of A over B^*. □

PROPOSITION 4. *Let $\mathbb{T}$ be a homogeneous 2-tournament with $T_1 \simeq \mathbb{Q}$, $T_2 \simeq \mathbb{Q}$ or $S(2)$, and suppose that for every $a \in T_1$ and every cross-type p, a^p is dense in T_2. Then $\mathbb{T} \simeq \Gamma_k(T_1, T_2)$, where k is the number of cross types.*

PROOF. Consider the set $\mathcal{A}$ of finite subtournaments T of T_2 such that: any 2-tournament of the form (L,T), with L finite and linear, embeds in $\mathbb{T}$. Our claim is simply that every tournament embedding in T_2 has this property. Consider also the following property, depending on a parameter n:

($*_n$) For some tournament $T(n)$ containing n disjoint copies of T, every 2-tournament of the form $(x, T(n))$ can be embedded in $\mathbb{T}$.

It is understood here that we refer to 2-tournaments whose cross types are taken from the supply of cross types in $\mathbb{T}$. A critical fact is that a tournament T which satisfies condition ($*_n$) for all n must be in our class $\mathcal{A}$. The argument for this is a completely typical application of an idea of Lachlan about the use of Ramsey's theorem in such situations. In order to keep the presentation reasonably self-contained, we shall sketch the proof of this fact afterwards. The method was introduced in [57], and is also described in [18].

Now it is clear that every finite subtournament T of T_2 has the property ($*_n$) for all n, by our density hypothesis, taking $T(n)$ to be any subtournament of T_2 containing n disjoint copies of T, for example the wreath product (composition) $T[L]$ where L is linear of order n. So every finite tournament embedding in T_2 lies in $\mathcal{A}$, as claimed. □

Now we shall sketch the necessary Ramsey argument. We remark that Lachlan's original argument took place in a substantially more complicated situation, but the argument itself was exactly as follows. We shall also need variants of this argument later, in more complicated situations, but as the argument itself never changes we shall not repeat the proof later on.

LEMMA 2.5. *Let $\mathbb{T}$ be a homogeneous 2-tournament with at least two cross types, and let L be a linear tournament of order n. There is a number N so that if $T, T(N)$ are tournaments for which:*

1. *$T(N)$ contains N disjoint copies of T;*
2. *Every 2-tournament of the form $(x, T(N))$ embeds in $\mathbb{T}$;*

then every 2-tournament of the form (L, T) also embeds in $\mathbb{T}$.

PROOF. Let k be chosen so large that any tournament on k vertices necessarily contains a linear order of size n. Let U be a set of k vertices $\{u_1, \ldots, u_k\}$, and let N_0 be the number of ways a linearly ordered sequence of n points can be selected from U, that is $N_0 = \binom{n}{k} \cdot n!$. Let $N = N_0 + k$.

Let $T(N)$ be chosen to satisfy (1,2), and let T_i be the $i - th$ copy of T in $T(N)$. We claim that there are 2-tournaments $\mathbb{A}_i = (\{u_i\}, T(N))$ such that any amalgam of the $\mathbb{A}_i$ over their common part $T(N)$ necessarily embeds (L, T). This will complete the argument: by (2) the 2-tournaments $\mathbb{A}_i$ all embed in $\mathbb{T}$, and by homogeneity some amalgam of them also embeds in $\mathbb{T}$.

So it remains only to define $\mathbb{A}_i$. Fix two distinct cross types p, q. For $i, j \leq k$ distinct let $T_{N_0+i} \subseteq u_i^p \cap u_j^q$. This ensures that no two of the vertices u_i can possibly become identified in the amalgam. Now enumerate the N_0 ordered sequences of length n of distinct elements which can be formed from the set U. If L_i is the ith such sequence, where $i \leq N_0$, connect the vertices in L_i to T_i in such a way that (L_i, T_i) will become isomorphic to (L, T) if L_i is given the structure of L. There are many edges in the $\mathbb{A}_i$ whose orientation has not yet been specified; these may be oriented arbitrarily.

The proof is over at this point. To recapitulate: after amalgamating the 2-tournaments $\mathbb{A}_i$ over $T(N)$, in the amalgam there will be at least one copy of L induced on the vertices in U; if this is the ith possible copy L_i, according to our initial listing of all the possibilities, then we discover that $(L_i, T_i) \simeq (L, T)$, as a result of our foresight. □

PROPOSITION 5. *Let $\mathbb{T}$ be a homogeneous 2-tournament with $T_1, T_2 \simeq S(2)$, and suppose that for every $a \in T_1$ and every cross-type p, a^p is dense in T_2. Then $\mathbb{T} \simeq \Gamma_k(T_1, T_2)$, where k is the number of cross types.*

PROOF. The proof of the previous proposition continues to work in this case, as far as it goes, producing the conclusion that any 2-tournament (L, T) with L linear, $T \leq S(2)$, and suitable cross types, does in fact embed in $\mathbb{T}$. This is not enough, but it will be useful for $|L| = 2$. Also notice that our density condition remains valid with the roles of T_1, T_2 interchanged, by Proposition 3.

We now fix a cross type p and an element $x \in T_1$, and consider the 3-tournament $\mathbb{T}^* = (x', {}'x, x^p)$ with components isomorphic to $\mathbb{Q}, \mathbb{Q}, S(2)$ respectively. Here (T_1^*, T_2^*) is shuffled with two cross types. The 2-tournaments (T_1^*, T_3^*) and (T_2^*, T_3^*) are both of the form $\Gamma_k(\mathbb{Q}, \mathbb{Q}^*)$, by the preceding proposition. The hypothesis of that proposition is satisfied by these 2-tournaments, in view of the remark in the preceding paragraph, with $|L| = 2$.

This is the situation described in Lemma 2.3. To apply that lemma, we only need to verify that $\mathbb{T}^*$ embeds all possible transversals. Embedding a transversal into $\mathbb{T}^*$ amounts to embedding a 2-tournament into $\mathbb{T}$ which is of the form $(A_1, \{b\})$ where $|A_1| \leq 3$. Our density condition allows this whenever A_1 is finite. □

We have proved so far that all 2-tournaments in which one component is finite, and all 2-tournaments in which neither component is isomorphic with T^∞, are found in our catalog. So far the proofs involve fairly general lines of argument. We complete the analysis of the restricted 2-tournaments in the next proposition, which involves more computation.

PROPOSITION 6. *Let $\mathbb{T}$ be a homogeneous 2-tournament with $T_1 \simeq \mathbb{Q}$ or $S(2)$, $T_2 \simeq T^\infty$, and with k cross types. Then $\mathbb{T} \simeq \Gamma_k(T_1, T_2)$.*

PROOF. Let P be the set of cross types in $\mathbb{T}$.

Let $\mathcal{A}$ be the set of finite tournaments A such that: every finite 2-tournament $\mathbb{S}$ with $S_1 \leq T_1$, $S_2 \simeq A$ and with cross types in P embeds into $\mathbb{T}$. Then we claim

$\mathcal{A}$ is an amalgamation class.

This sort of observation will be used repeatedly (its roots are in [57] and [61]), with some variations. Suppose on the contrary we have two tournaments T^1, T^2 belonging to $\mathcal{A}$ such that every possible amalgam of T^1, T^2 over their common part T^0 lies outside of $\mathcal{A}$; that is, for each of these finitely many possible amalgams T we can find a finite 2-tournament (S_T, T) not embedding in $\mathbb{T}$, with the usual restrictions: $S_T \leq T_1$ and the cross types are in P. The disjoint union S of all the S_T can be extended to a tournament embedding in T_1 (this is a useful property

2.2. RESTRICTED HOMOGENEOUS 2-TOURNAMENTS.

of T_1 which can be seen by inspection but is unfortunately not known to hold in any very general context). Now let (S, T^1), (S, T^2) be 2-tournaments with cross types in P such that the induced structure on (S_T, T^i) ($i = 1, 2$) agrees with the structure induced by (S_T, T). By hypothesis (S, T^1) and (S, T^2) both embed in $\mathbb{T}$, but no amalgam of them over (S, T^0) does; and this contradicts the homogeneity of $\mathbb{T}$.

The nontrivial property of $\mathcal{A}$, which will be demonstrated below, is that $[I_1, C_3]$ belongs to it (this denotes an oriented 3-cyle C_3 dominated by a single vertex I_1). Now the classification of amalgamation classes of finite tournaments is known; this is equivalent to the classification of homogeneous tournaments, and that classification was reviewed above. In particular, if an amalgamation class of finite tournaments contains the tournament $[I_1, C_3]$, then it contains all finite tournaments. Our proposition is obtained by applying this remark to $\mathcal{A}$.

So it is sufficient to prove that: any finite 2-tournament $\mathbb{S}$ with $S_1 \leq T_1$, $S_2 \simeq [I_1, C_3]$, and with cross types in P, can be embedded in $\mathbb{T}$. We shall arrive at this result gradually. We first prepare the ground further.

(∗) Let $A \subseteq T_2$ be finite, and p a cross type over A. (A realization of p is then of the form (x, A) with $x \in T_1$.) If p is realized in $\mathbb{T}$, then it is realized by a dense subset of T_1.

The proof is by induction on $|A|$, starting from the case $|A| = 1$, which is easy since $T_1 \not\simeq T^\infty \simeq T_2$. For $|A| > 1$ we let $A = A_0 \cup \{y\}$, and we set $p_0 = p \upharpoonright A_0$, $q = \mathrm{tp}(y/A_0)$, $\mathbb{T}^* = (^{p_0}A_0, {}^qA_0)$. The first component of $\mathbb{T}^*$ is dense in T_1 by induction hypothesis, and the second is isomorphic to T^∞ by inspection. It then follows that any type realized in T_1^* over y is realized by a dense subset of T_1^*; this is the case $|A| = 1$ again. So (∗) holds.

For $p \in P$ and $y \in T_2$ let $\mathbb{T}(p) = (^p y, y')$. Then $\mathbb{T}_i(p) \simeq T_i$ ($i = 1, 2$) and:

(1) $\mathbb{T}, \mathbb{T}(p)$ have the same cross types.

In other words, we claim that any triangle $(x, \{y, z\})$ whose cross types p, q belong to P can be embedded in $\mathbb{T}$. If $p = q$ this is easily seen, and otherwise we can use the amalgamation:

$$\begin{array}{|cc|} \hline x_2^\bullet & \circ y_2 \\ \uparrow & \\ x_1^\bullet & \circ y_1 \\ \hline \end{array} \qquad \begin{array}{l} \mathrm{tp}(x_i y_i) = p; \\ \mathrm{tp}(x_i y_j) = q \ (i \neq j). \end{array}$$

whose factors are afforded by the observation (∗). So (1) follows.

Now since $\mathbb{T}, \mathbb{T}(p)$ have the same data, this construction can be iterated and yields:

(2) Any finite 2-tournament $\mathbb{S}$ with $|S_1| = 1$, S_2 linear, and with its cross types in P, embeds in $\mathbb{T}$.

Then an application of (∗) gives:

(3) Any finite 2-tournament $\mathbb{S}$ with $S_1 \leq T_1$, S_2 linear, and with its cross types in P, embeds in $\mathbb{T}$.

Next we prove:

(4) Any finite 2-tournament $\mathbb{S}$ with $|S_1| = 1$, $S_2 \simeq C_3$, and with its cross types in P, embeds in $\mathbb{T}$.

For this we use the following sort of amalgamation diagram:

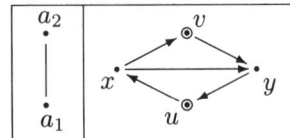

arranged so that upon insertion of an edge between u and v, we get either $(a_1, \{uxv\})$ or $(a_2, \{uvy\})$ isomorphic to $\mathbb{S}$. We shall make use of our freedom to select the orientation of the edge $a_1 a_2$ as well as the types $\text{tp}(a_1 y)$, $\text{tp}(a_2 x)$. In any case the factor omitting u is covered by (3), so we confine our attention to the factor omitting v.

Now there is no problem constructing suitable factors $(a_i, \{uxy\})$ by amalgamations determining $\text{tp}(a_1 y)$ and $\text{tp}(a_2 x)$, after which we determine the orientation of the edge $a_1 a_2$ by another amalgamation. This will fail if a_1, a_2 become identified, and to prevent this we shall take additional pains to ensure $\text{tp}(a_1/uxy) \neq \text{tp}(a_2/uxy)$. If $\mathbb{S}$ realizes 2 distinct cross types we can set the diagrams up so that $\text{tp}(a_1 u) \neq \text{tp}(a_2 u)$, and then there is no problem. So (4) holds in this case. If $\mathbb{S}$ realizes a unique cross type p and we arrive at the case $\text{tp}(a_1/uxy) = \text{tp}(a_2/uxy)$ then this forces $(a_1, \{uxy\}) \simeq \mathbb{S}$, so we can halt the construction. □

This completes the classification of restricted 2-tournaments, modulo Lemma 2.3, concerning 3-tournaments, which will be dealt with in due course.

2.3. Sources and sinks

To complete the classification of homogeneous 2-tournaments we still have to deal with 2-tournaments of type (T^∞, T^∞). Explicitly:

PROPOSITION 7. *Let $\mathbb{T}$ be a nondiagonal homogeneous 2-tournament with $T_1 \simeq T_2 \simeq T^\infty$. Let P be the set of cross types. Then there is a partition $\mathcal{P} = (P^0, P^+, P^-)$ of P with $P^0 \neq \emptyset$ such that $\mathbb{T} \simeq \Gamma(\mathcal{P})$ (up to a variant).*

We shall be occupied with the proof of this result for three more sections. In this section we show how to associate a suitable partition $\mathcal{P}$ with $\mathbb{T}$. In the following two sections we show that $\mathbb{T}$ is in fact the 2-tournament $\Gamma(\mathcal{P})$ canonically associated with $\mathcal{P}$. Roughly speaking, we deal first with the case $P^0 \neq P$, then with the case $P^0 = P$; but actually we shall carry out the whole argument by an induction which results in the two cases being interdependent.

As a matter of terminology, as soon as we have obtained the partition $\mathcal{P}$, we shall refer to the types in P^0, P^+, P^- as neutral types, sources, and sinks respectively. Recall that by definition, in $\Gamma(\mathcal{P})$ no triangle has an edge leading from a sink to a source.

We fix a nondiagonal homogeneous 2-tournament $\mathbb{T}$ with components T_1, T_2 isomorphic with T^∞. We define a relation R_1 on the set P of cross types by taking $R_1(p, q)$ to be true just in case there is no triangle with an edge in T_1 from q to p in the sense of the following diagram:

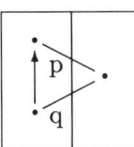

We define R_2 analogously, switching the roles of the two components. We let Γ_1 be the directed graph (P, R_1).

PROPOSITION 8. *There is a partition $\mathcal{P} = (P^0, P^+, P^-)$ of P, with $P^0 \neq \emptyset$, such that $R_1 = P^+ \times P^-$; and R_2 is either R_1, or its converse.*

PROOF. We shall divide the argument into five steps.

(1) Γ_1 is a directed graph with no loops or double edges

This is clear, since $\mathbb{T}$ is nondiagonal.

(2) Some vertex in P lies on no edge of Γ_1.

Suppose on the contrary that for every $p \in P$ we have a cross type $q = q(p)$ so that $R_1(p, q)$ or $R_1(q, p)$ holds. Fix $a \in T_2$ and choose a set $X = \{x_p : p \in P\} \subseteq T_1$ so that:

$$\text{tp}(x_p a) = q(p) \text{ for all } p \in P$$

Choose $b \in T_1$ so that $b \longrightarrow x_p$ iff $R_1(q(p), p)$. Observe that then $\text{tp}(ba)$ cannot be any type in P, a contradiction.

Now we prove a strange parallelogram law.

(3) If $p, q, r, s \in P$ and $R_1(p, q), R_1(r, s)$ both hold, then one of the following will hold:

$$R_2(p, s) \& R_2(r, q);$$
$$R_2(s, p) \& R_2(q, r).$$

We consider the following amalgamation, which exists in two variants, depending on the orientation of the edge in T_2.

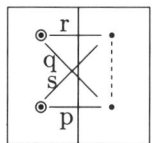

This diagram cannot be completed in $\mathbb{T}$. As $p \neq q$ and $r \neq s$, it is not possible to identify the circled vertices, and any orientation of an edge between them violates our hypothesis. Accordingly, one of the two factors does not embed in $\mathbb{T}$, and this occurs however the edge in T_2 is oriented. This yields the stated conclusion.

(4) No vertex of Γ_1 lies on both incoming and outgoing edges.

Indeed, just take $p = s$ in (3).

On the basis of (4), we can already define neutral types, sources, and sinks, and partition P accordingly into P^0, P^+, P^-, with P^0 nonempty.

(5) R_2 coincides with either R_1 or the converse of R_1

We use our parallelogram law. In the first place, taking $p = r$, $q = s$, every oriented edge in Γ_1 survives as an oriented edge in Γ_2, possibly with the opposite orientation.

Thus the neutral types in (P, R_2) must be neutral in (P, R_1), and hence by symmetry the two notions of neutrality coincide. Call a type in P^+ or P^- *positive* if it remains in the same category relative to the relation R_2, and *negative* otherwise. Now an edge in Γ_1 must connect two positive or two negative types, so we may also refer to the edges as positive or negative, accordingly. Suppose Γ_1 contains a positive edge (p, q) as well as a negative edge (r, s). Then we contradict (3). So in fact (5) holds.

Of course, as we may reverse the orientation of all the edges in T_2 if we wish, we may assume that $R_1 = R_2$, and write the parallelogram law (3) exclusively in terms of R_1. Thus if p is a source and s is a sink, we can find cross types q, r to which (3) applies, and conclude $R_1(p, s)$; that is, $R_1 = P^+ \times P^-$, as claimed. □

We always can and shall assume that the variant of $\mathbb{T}$ has been taken in which the two relations R_1, R_2 coincide. In particular the *constraint partition* (P^0, P^+, P^-) is unambiguously defined. As a matter of terminology, if $P^0 = P$ we say the partition is *degenerate*, or equivalently that $\mathbb{T}$ is *unconstrained*. In the next two sections we undertake an inductive proof that our nondiagonal tournament with components isomorphic to T^∞ is in fact $\Gamma_k(\mathcal{P})$. The induction is on the number k of cross types, and for fixed k the constrained case is treated prior to the unconstrained case. The two cases are treated separately in the following two sections.

2.4. Constrained 2-tournaments

The assumptions in force at present are:
(1) (T_1, T_2) is homogeneous with components $T_1 \simeq T_2 \simeq T^\infty$;
(2) $\mathbb{T}$ is nondiagonal;
(3) There are k cross types;
(4) $\mathbb{T}$ has a nondegenerate constraint partition $\mathcal{P} = (P^0, P^+, P^-)$;
(5) If the components of $\mathbb{T}$ are switched, the same constraint partition is obtained;
(6) If $\mathbb{T}^*$ is a homogeneous 2-tournament satisfying hypotheses (1,2), with fewer cross types, then $\mathbb{T}^*$ is $\Gamma(\mathcal{P}^*)$.

We must show that $\mathbb{T} \simeq \Gamma(\mathcal{P})$. We use the following sort of reduction. If p is a cross type in $\mathbb{T}$ let $\mathbb{T}(p)$ be the 2-tournament $(^p y, y')$; we suppress the dependence on y. Let $P(p)$ be the set of cross types realized in $\mathbb{T}(p)$, let $R_1(p), R_2(p)$ be the relations defined on $P(p)$ and let $\mathcal{P}(p)$ be the corresponding partition into (relative) neutral types, sources, and sinks. Actually, we still have to verify that $\mathbb{T}(p)$ falls under our standing hypotheses (1,2) and appropriate versions of (3,4). There are three points that require checking:

LEMMA 2.6. *With the hypotheses and notation above:*
1. *Both components of $\mathbb{T}(p)$ are isomorphic with T^∞.*
2. *For p not a sink we have $P(p) = P$. For p a sink we have $P(p) = P - P^+$.*
3. *$\mathbb{T}(p)$ is not a diagonal 2-tournament.*

PROOF. 1. The first component requires attention. First take $x \in T_1$, and form $\mathbb{S} = (x', x^p)$. We know that $S_1 \simeq T^\infty$, and p is still a cross type of $\mathbb{S}$. Now if $S_2 \not\simeq T^\infty$, then $\mathbb{S} \simeq \Gamma_l(T^\infty, S_2)$ for some l, and hence for $y \in S_2$ we find $^p y \cap S_1 \simeq T^\infty$, from which our claim follows for $\mathbb{T}$. Accordingly, we may suppose that $S_2 \simeq T^\infty$.

For $y \in S_2$, we want to show that $^p y \cap T_1$ embeds $[I_1, C_3]$, and for this it suffices to show that $A_y :=\, ^p y \cap S_1$ embeds C_3.

Certainly A_y is nonempty. If $|A_y| = 1$ we find that S_1, S_2 carry a canonical bijective correspondence. In other words, there is an x-definable bijection between x' and x^p in $\mathbb{T}$. If $x_1, x_2 \in T_1$ then we have two correspondences carrying $x_1^p \cap x_2^p$ into two $\{x_1, x_2\}$-definable sets A_1, A_2. Their composition is an x_1, x_2-definable bijection between the two sets A_1 and A_2, and everything here — x_1, x_2, A_1, A_2 —

lies in $T_1 \simeq T^\infty$. By inspection we must have $A_1 = A_2$ and the bijection is the identity. The meaning of this is that our x-definable bijections are independent of x (where defined). Thus their union is a 0-definable bijection between T_1 and T_2, and $\mathbb{T}$ is diagonal, a contradiction.

To conclude that A_y embeds C_3, we need only exclude the possibility that $A_y \simeq \mathbb{Q}$. In this case we know that $(A_y, y' \cap S_2) \simeq \Gamma_l(\mathbb{Q}, T^\infty)$ for some l. But this allows us to form the familiar sort of amalgamation diagram in $\mathbb{S}$ shown:

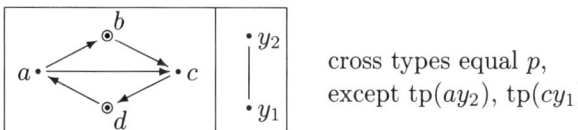

cross types equal p,
except $\mathrm{tp}(ay_2)$, $\mathrm{tp}(cy_1)$

We know that any diagram of this form with d omitted can be realized in $\mathbb{S}$, and even in $(A_y, y' \cap S_2)$. So the only difficulty is to determine the types of ay_2 and cy_1 so that the factor in which b is omitted can be seen to embed in $\mathbb{S}$. But for this just amalgamate factors (acd, y_1), (acd, y_2), each of which is constructed by an amalgamation which determines one of the types left unspecified. This completes the proof of (1).

We remark that we have gone through the part of the argument concerning the possibility that $|A_y| = 1$ in great detail because it is a typical instance of one which is needed quite often. We call this argument "variation of parameters", because it shows that a parametrically definable relation comes from a 0-definable relation, by studying the effect of varying the parameters and showing that any real dependence on these parameters would impose too much structure on a well understood situation. It may be objected that by homogeneity, since the language is binary there is a priori no way that one element x could be used to define a relation between two other elements, and indeed the only possibility is the one considered above: when restricted to certain x-definable sets, the binary relations already present may happen to acquire additional properties. Something of this sort happens in $\mathbb{S}(2)$, though not in a very pronounced manner.

2. Here there is nothing to prove; we merely record the facts.

3. This issue already arose in the proof of (1). It is handled by variation of parameters. □

LEMMA 2.7. *Let $p \in P^0 \cup P^+$. Then $\mathcal{P}_1(p) = \mathcal{P}_2(p) = \mathcal{P}$.*

PROOF. As $P(p) = P$, sources and sinks in P must remain sources and sinks in both $\mathcal{P}_1(p)$ and $\mathcal{P}_2(p)$, and hence these two partitions must coincide, since otherwise we know they would be dual. The point that requires proof, then, is that neutral types in $\mathbb{T}$ cannot become sources or sinks in $\mathbb{T}(p)$. In other words, we claim that if p, q, r are cross types and at least one of q or r is neutral in $\mathbb{T}$, then $\mathbb{T}$ embeds the configuration:

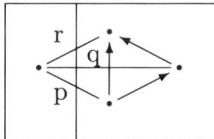

Fig. 1: pqr

In our discussion we shall use the symbols $+, -$ in our diagram to denote respectively some source and some sink. The whole discussion takes place in $\mathbb{T}$.

In the first place we suppose that q is neutral, and we notice that the following amalgamation diagram will produce a contradiction (assuming, that is, its factors are known to embed in $\mathbb{T}$):

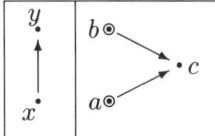

$\operatorname{tp}(x/abc) = pqr$;
$\operatorname{tp}(y/abc) = +-s$.

We first form the factor omitting a by an amalgamation over xb, which determines the type of yc. In this amalgamation, the factors will be triangles involving the neutral type q, so they do occur in $\mathbb{T}$. Observe that s is not a source, as $b \longrightarrow c$.

Once the type s has been determined, we need the factor omitting b. If s is a sink, then the orientation of the edge ac is forced on us, so our factor is obtained by amalgamating triangles known to occur in $\mathbb{T}$; here we should recall that p is not a sink. Suppose now that s is neutral. Then the following forced amalgamation with an auxiliary point o will be used (the orientation of oc needs to be determined):

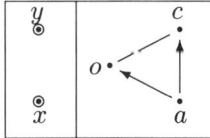

$\operatorname{tp}(x/aoc) = p + r$;
$\operatorname{tp}(y/aoc) = +-s$.

We still need to build up the factors of this amalgamation. These are produced as shown:

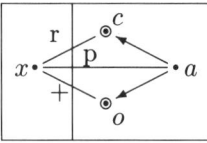 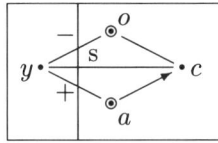

The lefthand amalgamation is performed first, determining the orientation of oc, and this orientation is fed into the second diagram. There are four triangles involved here. They all embed in $\mathbb{T}$ since s is neutral and p is not a sink. However if r is a source and the type denoted $+$ is r then the first amalgamation may not work, because the elements o and c may become identified. However in this case the configuration can be obtained by considering the structure of (x^r, x^p) which has factors of type T^∞ and which has only two cross types; hence the structure of this 2-tournament is known, by induction.

At this point we have proved that if q is not a sink in $\mathbb{T}$ then q cannot become a sink in $\mathbb{T}(p)$. Now we consider the second possibility, that the neutral type r becomes a source in $\mathbb{T}(p)$. Of course, if q is not a sink in $\mathbb{T}$, or equivalently in $\mathbb{T}(p)$, then our desired configuration automatically embeds in $\mathbb{T}$. So we now assume q is a sink in $\mathbb{T}$. Our contradiction in this case will come from the following amalgamation:

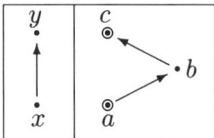

$\operatorname{tp}(x/abc) = pqr$;
$\operatorname{tp}(y/abc) = +s-$.

2.4. CONSTRAINED 2-TOURNAMENTS

To build up the factors in this amalgamation inside $\mathbb{T}$, we first construct the factor omitting a, determining the type s in the process. Since r is neutral this step is easy. Since q is a sink, the edge from x to y forces s to be a non-source. The second factor in our amalgamation is built using an auxiliary point:

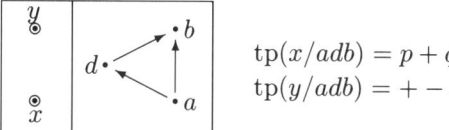

$$\operatorname{tp}(x/adb) = p + q;$$
$$\operatorname{tp}(y/adb) = + - s.$$

The edge from x to y is forced, but we still need to manufacture suitable factors for this last diagram. These are obtained from the following two amalgamations:

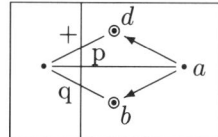 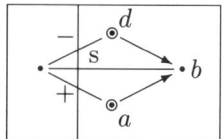

q is a sink, so both amalgamations admit unique solutions. The four factors occurring here are all triangles known to occur in $\mathbb{T}$, since p is not a sink and s is not a source. $\square$

LEMMA 2.8. *If $p \in P$ then $\mathcal{P}(p)$ is the restriction to $P(p)$ of $\mathcal{P}$, namely: $\mathcal{P}$ itself if p is not a sink, and degenerate otherwise.*

PROOF. In fact this has already been proved. Our claim is that any configuration of the type of Figure 1 will embed in $\mathbb{T}$ if all of the triangles involved in it do. By the previous lemma this holds unless p is a sink, and dually it also holds unless r is a source. If p is a sink and r is a source then the configuration does not even come into consideration. $\square$

LEMMA 2.9. *If $\mathbb{S}$ is any finite 2-tournament whose cross types all lie in $P^0 \cup P^-$, then $\mathbb{S}$ embeds in $\mathbb{T}$.*

PROOF. Let p be a sink. Our inductive hypothesis has been shown to be applicable to the 2-tournament $\mathbb{T}(p)$, which has fewer cross types, since $P(p) = P^0 \cup P^-$, and hence $\mathbb{S}$ is embedded in $\mathbb{T}(p)$, and thus certainly in $\mathbb{T}$. $\square$

After these preparations, we may turn to the proof that $\mathbb{T} \simeq \Gamma(\mathcal{P})$.

PROOF. We introduce the class $\mathcal{A}$ of all finite 2-subtournaments $\mathbb{S} = (S_1, S_2)$ of $\mathbb{T}$ for which we have:

- If $\mathbb{S}^* = (S_1 \cup L, S_2)$ is a linear extension of $\mathbb{S}$ (i.e., L is linear) with L realizing only sinks and neutral types over S_2, and if $\mathbb{S}^*$ embeds in $\Gamma(\mathcal{P})$, then $\mathbb{S}^*$ embeds into $\mathbb{T}$.

The embedding condition on $\mathbb{S}^*$ is just another way of saying that $\mathbb{S}$ respects the data in $\mathcal{P}$.

The class $\mathcal{A}$ is easily seen to be an amalgamation class. This is our second construction of this type, and they all follow the same principles. Here it is important for the amalgamation property that finitely many linear tournaments can be combined into one linear tournament. It is also important that we not allow both sinks and sources, or the amalgamation argument would fail.

Let $\mathbb{T}^*$ be the homogeneous 2-tournament which is associated with the amalagamation class $\mathcal{A}$, and similarly let P^* be its set of cross types, and let R_1^*, R_2^* be the edge relations in the associated graphs on P^*. Our claim is that $\mathbb{T}^*$ inherits the following properties from $\mathbb{T}$:

(1) $T_1^* \simeq T_2^* \simeq T^\infty$;
(2) $\mathbb{T}^*$ is nondiagonal;
(3) $P^* = P$;
(4) $R_1^* = R_2^* = R$.

Before verifying these claims, we should notice that they easily imply that $\mathbb{T} \simeq \Gamma(\mathcal{P})$. To see this, we argue that every finite 2-tournament $\mathbb{S}$ embedding in $\Gamma(\mathcal{P})$ must embed in every homogeneous 2-tournament satisfying (1 – 4). We proceed by induction on $|\mathbb{S}|$. If S_1 is empty then this is clear. Assume therefore that $\mathbb{S} = \mathbb{S}^0 \dot{\cup} \{x\}$ with $x \in S_1$.

Let $\mathbb{T}$ be any homogeneous 2-tournament satisfying (1-4). If x realizes both a source and a sink over S_2 then $\mathbb{S}$ is the unique solution to an amalgamation problem involving factors of smaller size, hence by induction these factors, and then also $\mathbb{S}$ itself, must embed in $\mathbb{T}$. In the remaining case, we may assume without loss of generality that x realizes only neutral types and sinks. Let $\mathbb{T}^*$ be the homogeneous 2-tournament associated with $\mathbb{T}$ as above. By induction, $\mathbb{S}^0$ embeds in $\mathbb{T}^*$, equivalently $\mathbb{S}^0 \in \mathcal{A}$. As $\{x\}$ is certainly linear, and realizes only neutral types and sinks over S_2, the definition of $\mathcal{A}$ guarantees that $\mathbb{S}$ embeds in $\mathbb{T}$.

It will be noticed that the foregoing argument is devoid of any real content, and that the actual proof of the lemma is therefore to be found in the verification of conditions (1-4). Furthermore, in the only application made of the definition of the class $\mathcal{A}$, only the case $|L| = 1$ is actually used; but as already noted, if we restrict the definition of $\mathcal{A}$ correspondingly, the whole proof collapses. These features are all present in the first application of these ideas by Lachlan in [57].

We turn to the verification of the properties (1–4). For T a tournament and $i = 1$ or 2, let T^i be the 2-tournament whose i^{th} component is T, and whose other component is empty. Observe that $T^1 \in \mathcal{A}$ for any finite tournament T^1, and thus $T_1^* \simeq T^\infty$. To prove (1) we must also see that for (L, T) finite with L linear, realizing only neutral types and sinks over T, we have an embedding of (L, T) in $\mathbb{T}$. This is contained in the previous lemma.

For the remaining three claims it suffices to show that any triangle of the following form which embeds in $\mathbb{T}$ is also in $\mathcal{A}$:

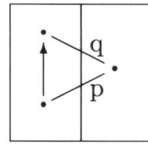

More generally, let $\mathbb{S}$ be a finite 2-tournament with $|S_2| = 1$ whose triangles embed in $\mathbb{T}$; we claim that $\mathbb{S}$ embeds in $\mathbb{T}$. This will be proved by induction on $|\mathbb{S}|$. Let $S_2 = \{x\}$. If x realizes both a source and a sink over S_1 our claim is immediate by induction via the corresponding amalgamation diagram. Otherwise, we may suppose that x realizes no source over S_1. The previous lemma then applies. □

2.5. Unconstrained 2-tournaments

At this point we have completed the classification of the constrained homogeneous 2-tournaments with k cross types, modulo the classification of the homogeneous 2-tournaments with fewer cross types.

2.5. Unconstrained 2-tournaments

The assumptions in force at present are:
(1) $\mathbb{T} = (T_1, T_2)$ is homogeneous with components $T_1 \simeq T_2 \simeq T^\infty$;
(2) $\mathbb{T}$ is nondiagonal;
(3) There are k cross types;
(4) $\mathbb{T}$ has a degenerate constraint partition $\mathcal{P} = (P, \emptyset, \emptyset)$;
(5) If $\mathbb{T}^*$ is a homogeneous 2-tournament satisfying hypotheses (1,2), with fewer cross types, then $\mathbb{T}^*$ is $\Gamma(\mathcal{P}^*)$;
(6) If $\mathbb{T}^*$ is a homogeneous 2-tournament satisfying hypotheses (1,2), with k cross types and a nondegenerate constraint partition, then $\mathbb{T}^*$ is $\Gamma(\mathcal{P}^*)$.

We must show that $\mathbb{T} \simeq \Gamma(\mathcal{P})$, as the final step in the classification of the homogeneous 2-tournaments (modulo Lemma 2.3). We use the tournaments $\mathbb{T}(p)$ which were introduced in the previous section. The proofs given there of the following facts continue to work in the present context:

LEMMA 2.10. *For $p \in P$:*
(7) $T_1(p) \simeq T_2(p) \simeq T^\infty$;
(8) $P(p) = P$;
(9) $\mathbb{T}(p)$ *is nondiagonal.*

We follow the same line of argument as in the previous section, especially as far as the final portion of the argument is concerned. We first prepare some lemmas relating to the structure of $\mathbb{T}(p)$.

LEMMA 2.11. *Any configuration of the form given in Fig. 1 (p. 29) embeds in $\mathbb{T}$ if two of the types p, q, r coincide.*

PROOF. If $q = r$ or $p = q$ apply (9) or its dual. If $p = r$ build the diagram:

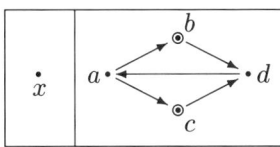

$\operatorname{tp}(x/abcd) = pqpp$.

We have the factor omitting b in $\mathbb{T}$, by the dual of (7), and we can get the factor omitting c from an amalgamation whose only other solution is the configuration we set out to build in the first place. □

LEMMA 2.12. *All types are neutral in $\mathbb{T}(p)$.*

PROOF. The argument used in the preceding section is not available here. Our claim is that any configuration of the form:

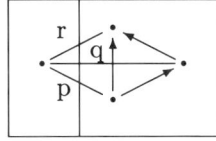

embeds in $\mathbb{T}$. We have to work a bit for this. Our aim is the amalgamation diagram:

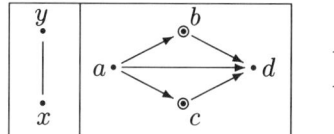

$\operatorname{tp}(x/abcd) = qqpr;$
$\operatorname{tp}(y/abcd) = pqrq.$

The problem is to find an orientation of the edge xy that lets us find both factors in $\mathbb{T}$. We introduce an auxiliary point o and we set up two amalgamation diagrams:

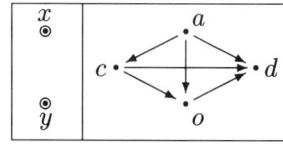

$\operatorname{tp}(x/acod) = qppr;$
$\operatorname{tp}(y/acod) = prpq.$

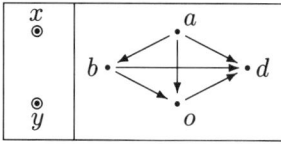

$\operatorname{tp}(x/abod) = qqpr;$
$\operatorname{tp}(y/abod) = pqpq.$

Now whenever either of these two diagrams is completed inside $\mathbb{T}$, we have in particular a triangle of the following form embedded in $\mathbb{T}(p)$:

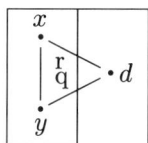

Now unless the pair $\{q, r\}$ consists of one source and one sink in $\mathbb{T}(p)$, there is no problem at all. On the other hand if we are dealing with a source and sink in $\mathbb{T}(p)$, then the orientation of the edge xy is fixed in both amalgamation diagrams, with the same orientation in both cases. So this orientation may be considered to be known.

What remains to be verified is that the four factors of our two amalgamation diagrams themselves embed in $\mathbb{T}$. These are all of the form:

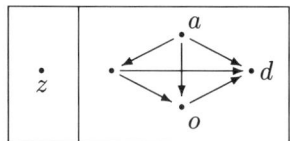

$\operatorname{tp}(z/a \cdot od) =:$
$qqpr;\ pqpq;\ qppr;$
or $prpq.$

In each case we can work in a derived 2-tournament of the form $\mathbb{T}(s) = (^s d, {}'d)$ with $s = r, q, r, q$ respectively. Now $P(s) = P$, and if $\mathbb{T}(s)$ is unconstrained then the previous lemma applies to $\mathbb{T}(s)$, while if $\mathbb{T}(s)$ is constrained then by (6) it is generic over a partition of P, so we need only check that $\mathbb{T}(s)$ embeds a suitable set of triangles. Expressed in terms of $\mathbb{T}$, what we have learned is that it suffices to examine all configurations of the form (z, X) with $X \subseteq \{a, \cdot, o, d\}$, $d \in X$, and $|X| = 3$.

Many of these are covered by the previous lemma. The remaining ones are all of the following form:

2.5. UNCONSTRAINED 2-TOURNAMENTS

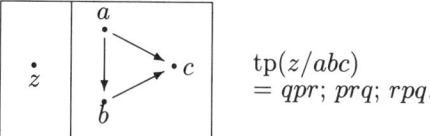
$\text{tp}(z/abc) = qpr;\ prq;\ rpq.$

The first two are handled by amalgamations with two solutions, one being the desired solution, while the other is the configuration we originally set out to construct. The amalgamation diagrams are:

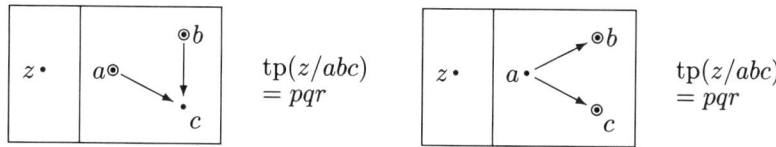

We are left with the third case to consider: $\text{tp}(z/a,b,c) = (r,p,q)$. Now it suffices to set up the diagram:

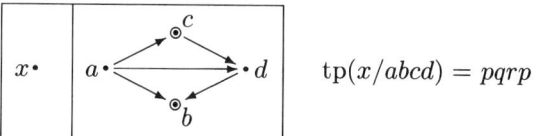

since this yields either the desired configuration or our initial goal. For the factors in this diagram the previous lemma suffices. □

LEMMA 2.13. *Any finite 2-tournament $\mathbb{L}$ with both components linear (and suitable cross types) embeds in $\mathbb{T}$.*

PROOF. If $|L_1| = 1$ we prove by induction on $|L_2|$ that $\mathbb{L}$ embeds in every unconstrained homogeneous 2-tournament with P as its set of cross types. For the inductive step it suffices to apply the previous step to a derived 2-tournament $\mathbb{T}(p)$.

Now the general case follows by Lachlan's Ramsey argument: force $\mathbb{L}$ to appear as the result of amalgamating several 2-tournaments of the form (x_i, L) where L contains many copies of L_2. □

LEMMA 2.14. *Any 2-tournament of the form $\mathbb{S} = (\{x\}, C_3)$ with suitable cross types can be embedded in $\mathbb{T}$.*

PROOF. Our usual diagram for this sort of claim is:

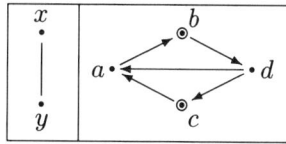

where one of the configurations $(x, \{cab\})$ or $(y, \{bdc\})$ will be isomorphic to $\mathbb{S}$. By the previous lemma we need only concern ourselves with the construction of the factor omitting c. The final step in the construction of that factor will be an amalgamation which determines the orientation of xy. (There is some slight worry that x, y might become identified, but this can be avoided unless we are working with a unique cross type over a, b, and c, in which case if x and y collapse then we already have the desired realization.)

So we are only concerned with embedding 2-tournaments of the following type in $\mathbb{T}$:

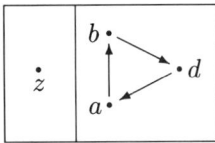

Here either $z = x$ and $\mathrm{tp}(zd)$ is immaterial, or $z = y$ and $\mathrm{tp}(zb)$ is immaterial. In either case an amalgamation of triangles will produce suitable configurations. □

LEMMA 2.15. *Let A be a copy of $[I_1, C_3]$ embedded in T_2. Then for any 1-type p over A made up of cross types in P, the derived 2-tournament $(^p A, A')$ embeds all triangles of the form:*

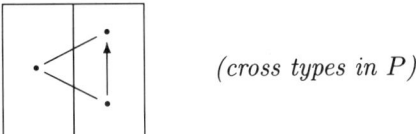

(cross types in P)

PROOF. This means that all cross 1-types over $[I_1, C_3, I_1, I_1] \leq T_2$ are realized in T_1. By passing to derived 2 tournaments defined successively from the three copies of I_1 named here, we obtain another unconstrained tournament $\mathbb{T}^*$ in which we need only realize a 1-type over C_3, and the previous lemma accomplishes this. □

We turn now to the proof that $\mathbb{T} \simeq \Gamma_k(T^\infty, T^\infty)$.

PROOF. We consider the amalgamation class $\mathcal{A}$ of finite 2-tournaments $\mathbb{S}$ such that any linear extension of $\mathbb{S}$ in the first component, with cross types in P, embeds in $\mathbb{T}$. As in the proof given in the previous section, we need only prove:

(10) For all finite tournaments T, we have $T^1, T^2 \in \mathcal{A}$;

(11) All triangles of the form $(x, \{a, b\})$ with cross types in P lie in $\mathcal{A}$.

Notice that (11) is essentially a special case of (10). The meaning of (11) is that tournaments of the form $(x \cup L, L_2)$, where L_2 is linear of order 2, embed in $\mathbb{T}$. Such a tournament may be thought of as a linear extension in the second component of a tournament of the form T^1. If we were to switch the roles of the two components in the definition of $\mathcal{A}$, this would then be covered by (10). So without loss of generality we may confine ourselves to the proof of (10), and since $\mathcal{A}$ is an amalgamation class of tournaments and these have been classified, we see that we only need to prove that $[I_1, C_3]^2 \in \mathcal{A}$.

Let $A = [I_1, C_3]$. We are trying to embed a 2-tournament of the form (L, A), with L linear, into $\mathbb{T}$. By Lachlan's Ramsey argument, we can reduce to this problem to a family of problems regarding the embedding of $(x_i, L_2[A])$ into $\mathbb{T}$, where L_2 is some much longer linear tournament. With a change of notation, therefore, our claim becomes:

- *Any cross 1-type p over $L[[I_1, C_3]]$ is realized in $\mathbb{T}$.*

We prove this by induction on $|L|$. For $|L| = 1$ it is the previous lemma. For $|L| > 1$ it follows from the induction hypothesis applied to the derived tournament $\mathbb{T}^* = (^{p_1} A, A')$ where A is the first copy of $[I_1, C_3]$ in $L[[I_1, C_3]]$ and $p_1 = p \upharpoonright$

A; assuming, that is, that this derived tournament is itself unconstrained, as the preceding lemma suggests. The only doubtful point is whether $T_1^* \simeq T^\infty$. If this is the case, we are done.

If $T_1^* \not\simeq T^\infty$ then since $T_2^* \simeq T^\infty$, $\mathbb{T}^*$ must be of the form $\Gamma_k(T_1, T^\infty)$. So all 1-types over $L[[I_1, C_3]]$ are realized in $\mathbb{T}^*$, hence also in $\mathbb{T}$. □

CHAPTER 3

Homogeneous n-tournaments

3.1. Introduction

Our goal is to show that any homogeneous n-tournament is uniquely determined by its 2-restrictions (the isomorphism type of each restriction to two components) and its transversals (n-subtournaments with at most one vertex in each component). We should notice that the classification of the homogeneous 2-tournaments is extremely simple, and that each one is uniquely determined by the isomorphism types of its components and the 2-tournaments it embeds which consist of a singleton and a linear component. So when we refer to the canonical data for a homogeneous n-tournament in the course of our analysis later on, one may think of this as referring to a list of of the transversals and the special 2-tournaments just described, together with a specification of the isomorphism type of each component.

Something more needs to be said about the 2-tournaments involved in the canonical data. For the most part these are triangles with one vertex in one component and two in the other. Unfortunately 2-tournaments with components of type S(2) behave differently. To determine the left/right partition in the event of shuffling (with the absence of shuffling corresponding to a degenerate partition) we need to know which 2-tournaments of the form (x, L) embed in $\mathbb{T}$ with L linear of order 3. If the three types realized by x over the points of L are p, q, r in order, the meaning of a failure of embeddability is that p, r are in one portion of the left/right partition, and q is in the other.

We shall call a finite n-tournament $\mathbb{S}$ "$\mathbb{T}$-restricted" if its canonical data are contained in those of $\mathbb{T}$. Our whole aim is to prove:

PROPOSITION 9. *If $\mathbb{T}$ is a homogeneous n-tournament and $\mathbb{S}$ is a finite $\mathbb{T}$-restricted n-tournament, then $\mathbb{S}$ embeds in $\mathbb{T}$.*

The proof of this should be considered as an induction on the total number of cross types occurring, though the inductive hypothesis is used very sparingly.

A finite 3-tournament $\mathbb{S}$ will be called *critical* if it has two degenerate components (singletons) and the third component consists of a pair of points which realize the same cross type over at least one of the two degenerate components. In the next two sections we prove a very special case of our proposition: we take $n = 3$ and $\mathbb{S}$ critical. However in the course of the proof we shall use the fact, proved below, that from the embedding of critical 3-tournaments slightly more general results follow immediately.

It turns out that much of the work and all of the really detailed amalgamation arguments appear in this very special case, or some closely related cases, after which the arguments become very smooth and natural.

3.2. Hypercritical and small 3-tournaments

$\mathbb{T}$ is a homogeneous 3-tournament throughout the present section. We shall assume that $\mathbb{T}$ has no finite component and no diagonal 2-restriction. Call a critical 3-tournament *hypercritical* if the two elements of the nondegenerate component realize the same cross type over each degenerate component. The next two lemmas concern aspects of the embedding problem for such 3-tournaments that can be handled without any very detailed analysis.

LEMMA 3.1. *Let $\mathbb{S}$ be a hypercritical $\mathbb{T}$-restricted 3-tournament. Then $\mathbb{S}$ embeds in $\mathbb{T}$.*

PROOF. Perhaps it suffices to say that this is proved by variation of parameters. In detail, let the nondegenerate component of $\mathbb{S}$ be S_1, and let p, q be respectively the cross type realized in S_1 over S_3, and the cross type realized in S_2 over S_3. If $\mathbb{S}$ is $\mathbb{T}$-restricted but omitted by $\mathbb{T}$, this means that for $x \in T_3$, the 2-tournament $(^p x, {}^q x)$ is diagonal with respect to the cross type realized in (S_1, S_2). For two choices of x_1, x_2 in S_3, this gives an $\{x_1, x_2\}$-definable bijection between the realizations of two types over $\{x_1, x_2\}$ in T_1. To understand what the possibilities are it suffices to examine the structure of (T_1, T_3), remembering that both components are infinite and that this 2-restriction is not diagonal. In fact there are no possibilities other than the identity. This means that the parameters vanish, and (T_1, T_2) is already diagonal, a contradiction. □

We shall deal with the critical case in the next section. More generally, call a 3-tournament *small* if it contains at most 4 vertices. We need to deal with small tournaments, but we shall see now that this reduces quickly to the critical case.

The following terminology will be useful. If $\mathbb{T}$ is a homogeneous 3-tournament, then a *derived* 3-tournament is the structure induced on any three distinct types realized in $\mathbb{T}$ over some single finite set $A \subseteq \bigcup_i T_i$. $\mathbb{T}$ is itself counted as one of its derived 3-tournaments.

LEMMA 3.2. *Let $\mathbb{T}$ be a homogeneous 3-tournament such that each of its derived 3-tournaments $\mathbb{T}^*$ satisfies:*

(∗) *]Every critical $\mathbb{T}^*$-restricted finite 3-tournament embeds into $\mathbb{T}^*$.*

Then every small $\mathbb{T}$-restricted 3-tournament embeds into $\mathbb{T}$.

PROOF. Let $\mathbb{S}$ be a small $\mathbb{T}$-restricted 3-tournament. The only case of interest is that in which $\mathbb{S}$ has a unique component with two elements. Let us suppose $S_1 = \{x\}$, $S_2 = \{y\}$, $S_3 = \{a, b\}$. We may suppose that none of the 2-restrictions of $\mathbb{T}$ is diagonal, and that a, b realize distinct types over x. We shall set up the following amalgamation:

$$\begin{array}{|c|c|c|} \hline x_2 & y_2 & z_2 \\ \uparrow & \uparrow & \\ x_1 & y_1 & z_1 \\ \hline \end{array} \qquad \mathrm{tp}(x_i y_j) = \mathrm{tp}(xy)$$

which should be arranged so that an edge in either direction between z_1, z_2 will produce as a copy of $\mathbb{S}$ either $x_1 y_1 z_1 z_2$ or $x_2 y_2 z_1 z_2$. The 2-types involved prevent the identificaton of z_1 and z_2.

The factors of this amalgamation diagram are of the form:

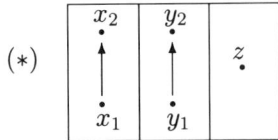

where x_1, x_2 realize the same type over y_1, y_2. Accordingly if we fix y_1 it suffices to embed the rest of this diagram into a suitable derived 3-tournament over y_1. But the rest of this diagram is a critical 3-tournament, so by our hypothesis it suffices to treat the configurations arising from (*) by omitting one of x_1, x_2, y_2, z. The first three are critical and $\mathbb{T}$-restricted, hence assumed to embed in $\mathbb{T}$. The last one is a 2-tournament which embeds in $\mathbb{T}$ because the 2-restrictions of $\mathbb{T}$ are assumed to be nondiagonal. □

3.3. The critical case

$\mathbb{T}$ is a homogeneous 3-tournament throughout the present section.

With the preliminaries of the previous section out of the way, we now fix a critical $\mathbb{T}$-restricted 3-tournament $\mathbb{S}$, which we shall embed in $\mathbb{T}$. We shall assume $\mathbb{S}$ has the form $(x, y, \{a, b\})$ with $a \longrightarrow b$, and label the cross types as follows:

$$\operatorname{tp}(xy) = s; \operatorname{tp}(xa) = \operatorname{tp}(xb) = r; \operatorname{tp}(y/a, b) = p, q \text{ respectively.}$$

For $x \in T_1$ we let $\mathbb{T}(x) = (x^s, x^r)$ (so e.g. $T_1(x) \subseteq T_2$). Our claim is that the following configuration embeds in $\mathbb{T}(x)$:

(C1)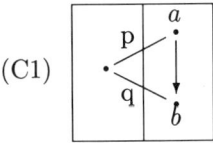

We may suppose that the types p, q are distinct, since otherwise the problem was treated in the previous section. The classification of the homogeneous 2-tournaments applies, and the claim is immediate unless $\mathbb{T}(x)$ falls under one of the two cases in which there are restrictions on such triangles, namely: $\mathbb{T}(x)$ shuffled with components of type $\mathbb{Q}$, or $\mathcal{P}$-generic for a nontrivial partition of the cross types. The shuffled case is fairly straightforward.

LEMMA 3.3. *If $\mathbb{T}(x)$ is shuffled for $x \in T_1$, then $\mathbb{S}$ embeds in $\mathbb{T}$.*

PROOF. We proceed by variation of parameters. There is an x-definable bijection between the dedekind completions of $T_1(x)$ and $T_2(x)$. For $x_1, x_2 \in T_1$, the set $A = T_1(x_1) \cap T_1(x_2)$ is dense in its convex closure in both $T_1(x_1)$ and $T_1(x_2)$, which is a cut (possibly improper) in both. Thus we have x_i-definable bijections f_i of the dedekind completion of A with the dedekind completions of two other $\{x_1, x_2\}$-definable sets B, C. This produces an $\{x_1, x_2\}$-definable bijection f of the dedekind completions of B and C. This situation exists in (T_1, T_3). The only possibility is that f is the identity. Accordingly the maps f_1, f_2 agree on their common domain and hence there is a global shuffle of (T_2, T_3) inducing the shuffle on $\mathbb{T}(x)$. As $\mathbb{S}$ is $\mathbb{T}$-restricted, the triangle above embeds in $\mathbb{T}$, and its cross types embed in $\mathbb{T}(x)$, so the triangle itself embeds in $\mathbb{T}(x)$. □

LEMMA 3.4. *Assume that $T_2 \simeq T_3 \simeq T^\infty$. Then $\mathbb{S}$ embeds in $\mathbb{T}$.*

PROOF. Another way to formulate the matter is: if p,q are not already respectively a sink and a source in $\mathbb{T}$, can this occur in $\mathbb{T}(x)$? We shall assume toward a contradiction that this does happen. In any case, passing to a variant of $\mathbb{T}$ if necessary, we may suppose that the sources and sinks do not change if we switch the components of $\mathbb{T}(x)$. Our goal is to build the following amalgamation diagram in $\mathbb{T}$:

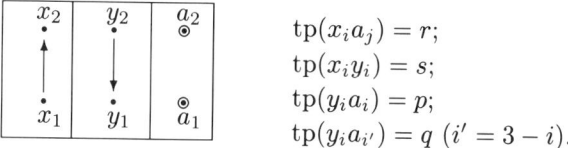

$$\mathrm{tp}(x_i a_j) = r;$$
$$\mathrm{tp}(x_i y_i) = s;$$
$$\mathrm{tp}(y_i a_i) = p;$$
$$\mathrm{tp}(y_i a_{i'}) = q \ (i' = 3 - i).$$

The two factors have a common form:

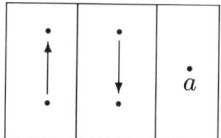

If we fix the parameter a, then we are trying to embed the following two small 3-tournaments in a derived 3-tournament $\mathbb{T}[a] = (^r a, {}^p a, {}^q a)$:

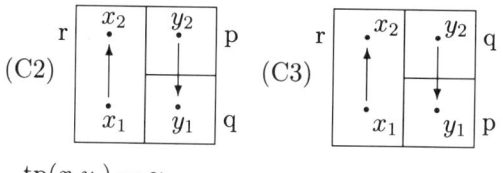

$\mathrm{tp}(x_i y_i) = s;$
the types $u = \mathrm{tp}(x_1 y_2)$ and $v = \mathrm{tp}(x_2 y_1)$
both remain to be determined.

We distinguish two cases, depending on the isomorphism type of T_1.

Case 1. $T_1 \not\cong T^\infty$.

We claim first that Lemma 3.2 applies to $\mathbb{T}[a]$. Let $\mathbb{T}^*$ be a derived 3-tournament associated with $\mathbb{T}[a]$. In any 2-restriction of $\mathbb{T}^*$ with both components of type T^∞ there will be only two cross-types. So this 2-restriction will not be diagonal, and since there must be a neutral type it follows that the 2-restriction is unconstrained. We have already completed the proof of the embedding of critical $\mathbb{T}^*$-restricted 3-tournaments in such cases, since in fact the only case not yet treated is the subject of the present lemma, and involves sources and sinks in a derived 3-tournament.

So by Lemma 3.2, the configurations above can be embedded in $\mathbb{T}[a]$ if their proper subconfigurations can be. There are eight of these. Four of them (omitting some y_i) are trivial, since $T_1[a]$ is not isomorphic to either of the other two components, and hence there can be no nontrivial constraints on triangles, by the 2-tournament classification. (In other words, for these four configurations it suffices to know that the transversals of $\mathbb{S}$ embed in $\mathbb{T}$.)

We are left with four configurations in which the as yet undetermined types u, v occur, and the problem is to choose these two types so that all four of the configurations occur in $\mathbb{T}[a]$:

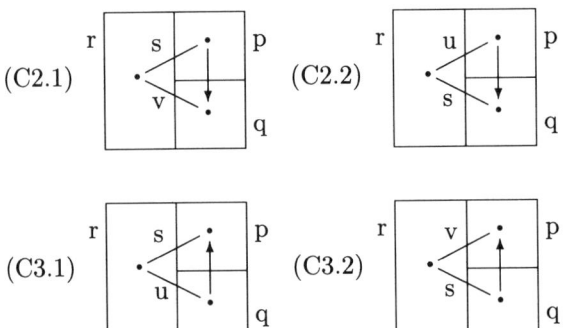

We have two configurations involving u and two involving v, and the appropriate choice of u, v involves detailed amalgamation arguments. As an example we shall discuss the choice of v.

The discussion takes place once more in $\mathbb{T}$. The first step is to build (C2.1) by an amalgamation which determines v. In $\mathbb{T}$ the diagram is as follows:

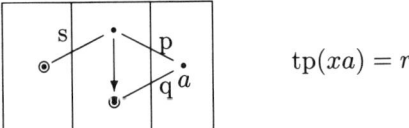

$$\operatorname{tp}(xa) = r$$

One factor is a transversal of $\mathbb{S}$, hence embeds in $\mathbb{T}$. The other factor is a triangle whose failure to embed in $\mathbb{T}$ would make the embedding problem for $\mathbb{S}$ trivial; so we assume it embeds in $\mathbb{T}$. Note that v cannot be s since we assumed that p is a sink and q a source in $\mathbb{T}(x)$.

Now that the type v is chosen, we need to build the companion factor (C3.2). Our first attempt is as follows.

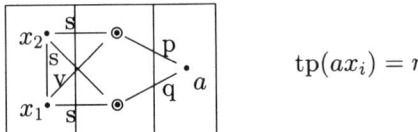

$$\operatorname{tp}(ax_i) = r$$

We are using the hypothesis that p, q become respectively a sink and a source in $T(x_2)$ to determine the orientation of the inserted edge. The question of course is whether the factors of this diagram are to be found in $\mathbb{T}$:

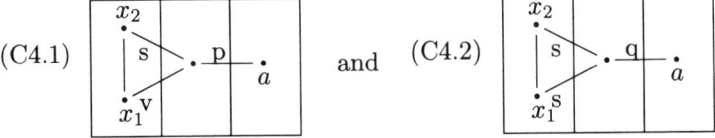

Here $\operatorname{tp}(x_i a)$ should be r for $i = 1, 2$, and we have to fix an orientation of the edge $x_1 x_2$.

Now there is no difficulty with the second of these factors, as the relevant two components of $\mathbb{T}[a]$ are not isomorphic, hence not diagonal. Therefore we concern ourselves only with the first factor, which is naturally manufactured by an amalgamation which determines the orientation of the edge $x_1 x_2$, and since v, s are distinct no collapse of two elements to one is possible. So the whole problem reduces to the two transversals needed to form this amalgamation, one of which occurs in $\mathbb{S}$, while the other is:

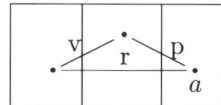

If this particular transversal happens to be omitted by $\mathbb{T}$, we argue in a different manner. Fix $a \in T_3$ and consider the derived 3-tournament $\mathbb{T}^*(a) = (^r a, {}^p a, {}'a)$. Since we are assuming this omits the cross type v, we may finally invoke the inductive hypothesis to conclude that $\mathbb{T}^*(a)$ will embed $\mathbb{S}$ if it embeds all of the proper subconfigurations of $\mathbb{S}$. In terms of $\mathbb{T}$ we are talking about four configurations, each with four vertices. Two of them are 2-tournaments, one is hypercritical, and the remaining one is:

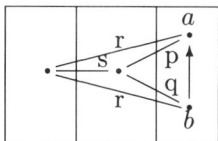

which is the original configuration with the orientation of ab reversed. So if the original configuration is omitted, then this one is embedded, and we reach a contradiction.

Case 2. $T_1 \simeq T^\infty$.

This is the most troublesome case, and requires a further case division. We shall say that the type s is a source over q, r if it becomes a source in $(^r x, {}^q x)$ for $x \in T_3$. We may assume that sources and sinks do not change if we switch the two components.

Case 2.1. s is a source over q, r.

For $a \in T_3$ let $\mathbb{T}[a]$ be $(^r a, {}^q a, a')$. Observe that, like $\mathbb{T}$, $\mathbb{T}[a]$ also embeds the proper subconfigurations of $\mathbb{S}$, and that $\mathbb{T}[a]$ also has all its components isomorphic with T^∞. Accordingly we may replace $\mathbb{T}$ by $\mathbb{T}[a]$. As a result we may now suppose that s is a source in $\mathbb{T}$ itself. We are going to aim at the following configuration, equivalent to $\mathbb{S}$ since we have assumed that sources and sinks in $\mathbb{T}(x)$ are unchanged after switching components:

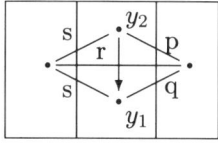

We aim at the following amalgamation diagram:

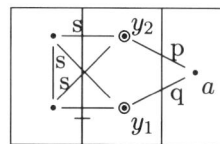

The factor omitting y_1 is hypercritical, hence embeds in $\mathbb{T}$. The other factor may be obtained by amalgamating its two transversals, one of which has been assumed to embed in $\mathbb{T}$, while the other is:

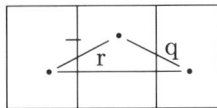

If the latter is omitted we can fix $a \in T_3$ and argue inductively inside $(^r a, {}^P a, {}' a)$. This completes our analysis when s is a source over q, r. A similar analysis applies if s is a sink over p, r.

Case 2.2. s is neither a source over q, r, nor a sink over p, r.

For $a \in T_3$ we again use the notation $\mathbb{T}[a]$ for $(^r a, {}^P a, {}' a)$. Observe that $\mathbb{S}$ is $\mathbb{T}[a]$-restricted, making use of Lemma 3.1 for one transversal, and a simple amalgamation argument for the other. In particular we may suppose that $\mathbb{T}[a]$ realizes the same types as $\mathbb{T}$, and has the same partition of types in each 2-restriction as $\mathbb{T}$; otherwise we can argue inductively on the number of types, and then also on the number of neutral types. We can make the same hypothesis for the dual construction $\mathbb{T}[a] = (^r a, {}^q a, a')$.

In this case we return to the method of Case 1. We refer again to $\mathbb{T}[a]$ and to the configurations (C2,C3) used in that case.

We shall now manufacture (C2), which in $\mathbb{T}$ is:

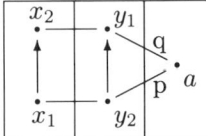

$\mathrm{tp}(x_i y_i) = s;$
$\mathrm{tp}(x_i a) = r;$
$\mathrm{tp}(x_1 y_2) = u;$
$\mathrm{tp}(x_2 y_1) = v.$

We first determine u by amalgamating $x_1 y_1 a$ and $y_1 y_2 a$ over $y_1 a$. Then we determine v by in (C2) by an amalgamation, whose first factor we have just constructed, while the second is:

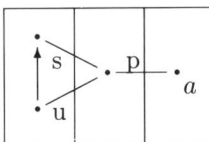

$\mathrm{tp}(x_i a) = r.$

If this configuration does not embed in $\mathbb{T}$, we are just saying that u is a sink and s is a source in $(T_1[a], T_2[a])$. But by our initial reduction, this means that u is a sink and s is a source in (T_1, T_2). This contradicts the construction of u. So the second factor is indeed available, and we have embedded (C2) into $\mathbb{T}$, determining the types u, v in the process. It remains to embed the following configuration into $\mathbb{T}$:

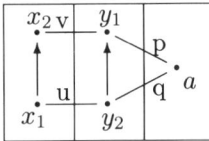

$\mathrm{tp}(x_i y_i) = s;$
$\mathrm{tp}(x_i a) = r.$

The amalgamation diagram used for this purpose will be:

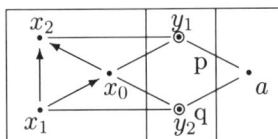

$\mathrm{tp}(x_i a) = r;$
$\mathrm{tp}(x_i y_j) = \mathrm{s}$ except for:
$\mathrm{tp}(x_1 y_2) = \mathrm{u}; \mathrm{tp}(x_2 y_1) = \mathrm{v}.$

Consider its factors:

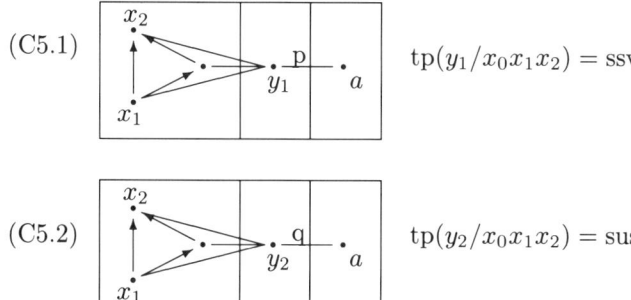

with $\mathrm{tp}(x_i a) = r$ in both.

Working in the corresponding derived 2-tournaments over a, we see that it suffices to check that they are not s-diagonal, which is ensured by Lemma 3.1, and that they embed certain triangles corresponding to these two factors with x_0 omitted. By our present case assumption – s is not a sink over (p, r), nor a source over (q, r) – this will hold if $\mathbb{T}$ embeds the transversals:

By our initial reduction, we assume the cross types u, v are not lost in going to either of the two 3-tournaments we have associated with a, so we have these transversals in $\mathbb{T}$.

$\square$

COROLLARY 2. *Let $\mathbb{T}$ be a homogeneous 3-tournament, $\mathbb{S}$ a small $\mathbb{T}$-restricted 3-tournament. Then $\mathbb{S}$ embeds in $\mathbb{T}$.*

PROOF. We combine Lemma 3.2 with the results of this section. $\square$

We seem to have worked rather hard for very little information, but as we indicated at the outset the complete classification of the homogeneous n-tournaments will soon be gotten quite easily, and in a natural way. As a first step, we shall convert the previous result into something really useful in the next section.

3.4. Two embedding lemmas.

In this section we give natural extensions of our embedding lemma for 3-tournaments to similar classes of n-tournaments. In this section $\mathbb{T}$ is a homogeneous n-tournament. We shall have to return to the case of 3-tournaments once more in the following section.

We remind the reader that the canonical data associated with $\mathbb{T}$ may be thought of either as a list of the transversals and 2-restrictions of $\mathbb{T}$, or – more concretely – as a list of the transversals and 2-tournaments (x, L) embedding in various 2-restrictions of $\mathbb{T}$ with L linear of order at most 3, together with the isomorphism types of the individual components. One could also associate extremely simple invariants with these isomorphism types, but this is not very useful in general. On the other hand we shall pay close attention to transversals and triangles.

LEMMA 3.5. *Let $\mathbb{S}$ be a finite $\mathbb{T}$-restricted n-tournament whose components are linear, with $|S_n| = 1$, and whose 2-restrictions (S_i, S_j) each have a unique cross type for $i, j < n$. Then $\mathbb{S}$ embeds in $\mathbb{T}$.*

PROOF. We may suppose $n > 2$. We proceed by induction on n and on $|\mathbb{S}|$. Fix some $i < n$ and let a be the first element of S_i. Then there is an associated derived n-tournament $\mathbb{T}(a)$ into which we wish to embed $\mathbb{S} - \{a\}$. By induction it suffices to check that $\mathbb{S} - \{a\}$ is $\mathbb{T}(a)$-restricted. Equivalently, for $\mathbb{U}$ either a component, a 2-tournament with one component of order 1 and the other component linear, or a transversal of $\mathbb{S}$, we have to check that $\mathbb{U} \cup \{a\}$ embeds in $\mathbb{T}$. But this follows by induction unless $\mathbb{U} \cup \{a\} = \mathbb{S}$. If $|S_i| \geq 3$ there can be no problem, so we may assume that no component of $\mathbb{S}$ has more than two elements. Since $n > 2$, the only possibilities for $\mathbb{U}$ with $\mathbb{U} \cup \{a\} = \mathbb{S}$ are:

(1) $\mathbb{U}$ is a triangle meeting two components of $\mathbb{S}$, $n = 3$, and $\mathbb{S}$ is critical;
(2) $\mathbb{U}$ is a transversal, $|\mathbb{S}_i| \leq 1$ with one exception, of order 2.

The critical case, and more generally the small case, was disposed of in the previous section. So case (1) does not come into consideration, nor does case (2) if $n = 3$. So we are in case (2), with $n > 3$, and the problem which arises in that case is avoided by choosing i so that $|S_i| = 1$. □

LEMMA 3.6. *Let $\mathbb{S}$ be a $\mathbb{T}$-restricted n-tournament with $|S_i| = 1$ for $i < n$, and with $|S_n|$ linear. Then $\mathbb{S}$ embeds in $\mathbb{T}$.*

PROOF. We proceed by induction on $|S_n|$. The case $|S_n| = 1$ is trivial. If $|S_n| > 1$ let a be the first element of S_n. Let $\mathbb{T}(a)$ be the appropriate derived n-tournament whose i^{th} component is given by the type of S_i over a for $i < n$, and whose last component is a'. So our claim is that $\mathbb{S} - \{a\}$ embeds in $\mathbb{T}(a)$, or using the induction hypothesis, simply that $\mathbb{S} - \{a\}$ is $\mathbb{T}(a)$-restricted. So we have two claims:

(1) For any 2-restriction $\mathbb{U}$ of $\mathbb{S} - \{a\}$, $\mathbb{U} \cup \{a\}$ embeds in $\mathbb{T}$;
(2) For any transversal $\mathbb{U}$ of $\mathbb{S} - \{a\}$, $\mathbb{U} \cup \{a\}$ embeds in $\mathbb{T}(a)$.

Now with regard to (1) it suffices to notice that for $\mathbb{U}$ a 2-restriction of $\mathbb{S} - \{a\}$, $\mathbb{U} \cup \{a\}$ is either a transversal or 2-restriction of $\mathbb{S}$, hence embeds in $\mathbb{T}$ by hypothesis.

Our second claim is simply that the lemma holds if $|S_n| = 2$. We already know this if $n \leq 3$, so assume $n > 3$. Let $\mathbb{A} = \mathbb{S} - S_n = \{a_i : i < n\}$ and take two copies $\mathbb{A}^j = \{a_i^j\}$ ($j = 1, 2$) of $\mathbb{A}$ with:

$$\text{tp}(a_i^j a_{i'}^{j'}) = \text{tp}(a_i a_{i'}) \quad \text{if } i \neq i'; \qquad a_i^1 \longrightarrow a_i^2;$$

and form an amalgamation diagram:

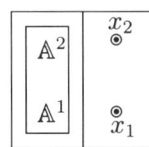

in such a way that one of $\mathbb{A}^1 x_1 x_2$, $\mathbb{A}^2 x_1 x_2$ will become isomorphic with $\mathbb{S}$ however the edge $x_1 x_2$ is oriented. Of course if x_1, x_2 realize the same type over $\mathbb{A}^1 \cup \mathbb{A}^2$ then they might become identified, so we argue differently in this case, reducing to Lemma 3.1 by an easy induction, using a derived $(n-1)$-tournament.

Our problem now is to check that the factors $\mathbb{A}^1 \mathbb{A}^2 x_i$ both embed in $\mathbb{T}$. This is the content of the previous lemma. □

3.5. Polarized n-tournaments

For the remainder of the chapter, $\mathbb{T}$ will be a homogeneous n-tournament. Our ultimate goal is to prove that an arbitrary finite $\mathbb{T}$-restricted n-tournament will embed in $\mathbb{T}$, and hence that $\mathbb{T}$ is determined by its canonical data, though we shall limit ourselves to slight extensions of the embedding lemma for small 3-tournaments up to the end of the next section. We may impose the following conditions on $\mathbb{T}$ with no loss of generality:

(1) $\mathbb{T}$ has no diagonal 2-restrictions;
(2) All components of $\mathbb{T}$ are infinite.

Indeed, if $\mathbb{T}$ has a diagonal 2-restriction, then it is uniquely determined by one of its $(n-1)$-restrictions, the diagonal 2-restriction, and its transversals of order 3. So this allows the first reduction to be made. For (2), let $\mathbb{T}^f$ be the union of the finite components of T. As the components of $\mathbb{T} - \mathbb{T}^f$ are infinite and primitive, each of them realizes a single type over $\mathbb{T}^f$. So the structure of $\mathbb{T}$ is determined by the structure of $\mathbb{T}^f$, of $\mathbb{T} - \mathbb{T}^f$, and the 2-restrictions involving one infinite and one finite component. The only point that needs to be checked is that $\mathbb{T}^f$ is itself determined by its canonical data. This is not really essential, since we would have no objection to including the structure of $\mathbb{T}^f$ in the canonical data, but as it happens $\mathbb{T}^f$ is necessarily very simple: the components are singletons I_1 or 3-cycles C_3, and after the reduction (1), any 2-restriction has a unique cross type. So the 2-restrictions suffice to determine $\mathbb{T}^f$ in this case.

By the classification of 2-tournaments, condition (2) will also apply to any derived tournament associated with $\mathbb{T}$. We do not wish to examine (1) in detail at this time, but it is worth noting that this condition will be inherited by any derived tournament defined over a single point. This is proved by variation of parameters.

We now modify the "source/sink" terminology so as to be able to discuss the shuffled and $\mathcal{P}$-generic cases in a uniform way. We call a cross type p occurring in the 2-restriction (T_i, T_j) of $\mathbb{T}$ an (i,j)-*source* in either of the following two cases: (T_i, T_j) is $\mathcal{P}$-generic and p is a source in the usual sense, or (T_i, T_j) is shuffled and p belongs to the L-set of types. The notion of (i,j)-*sink* is defined dually. As we have seen, in shuffled 2-tournaments the sources and sinks partition the cross-types, while in the $\mathcal{P}$-generic setting there will also be neutral types. We have also seen that interchanging the two components results in the same partition, but the rôles of the sources and sinks may be reversed. Sources and sinks behave in a somewhat troublesome manner in shuffled 2-tournaments with components of type S(2). They do not involve the exclusion of triangles, but they do limit the 1-types realized over linear orderings of length at least 3.

Let $\tau : \{1, 2, \ldots, n-1\} \to \{\pm 1\}$. We shall call an n-tournament $\mathbb{S}$ τ-*polarized* if for every $i < n$ we have:

$$\begin{aligned} \text{if } \tau(i) = 1 &\quad \text{then } (S_i, S_n) \text{ realizes no } (i, n)\text{-sink}; \\ \text{if } \tau(i) = -1 &\quad \text{then } (S_i, S_n) \text{ realizes no } (i, n)\text{-source}. \end{aligned}$$

This definition depends on $\mathbb{T}$, from which the notions of sink and source are derived. It also depends on n, which will sometimes need to be replaced by another index i. In this case we speak of polarization *at i*, or *at the i-th component*.

With $\mathbb{T}$ and τ both fixed, consider the class $\mathcal{A}(\tau)$ of finite $\mathbb{T}$-restricted $(n-1)$-tournaments $\mathbb{A} = (A_1, \ldots, A_{n-1})$ such that:

- *Every finite linear extension* $\mathbb{S} = (\mathbb{A}, L)$ *of* $\mathbb{A}$ *to a τ-polarized $\mathbb{T}$-restricted n-tournament embeds in* $\mathbb{T}$.

Here $(\mathbb{A}, L)$ means $(A_1, \ldots, A_{n-1}, L)$.

LEMMA 3.7. $\mathcal{A}(\tau)$ *is an amalgamation class.*

PROOF. We have seen a few examples of this type of argument. The main point is that if $L_1, \ldots, L_k$ are finite linear tournaments and all the $(\mathbb{A}, L_i)$ are $\mathbb{T}$-restricted τ-polarized n-tournaments, then the extension $(\mathbb{A}, L)$ with $L = [L_1, \ldots, L_k]$ (linear) is again $\mathbb{T}$-restricted and τ-polarized. (The reductions (1,2) above both play a rôle here.) The rest is as in earlier examples – from a counterexample to amalgamation in $\mathbb{A}(\tau)$ concoct a counterexample to amalgamation in the substructures of $\mathbb{T}$. □

3.6. Embedding polarized 3-tournaments

In the preceding section we showed how the notion of polarization in n-tournaments leads to natural amalgamation classes of $(n-1)$-tournaments. If we aim to classify the homogeneous n-tournaments by induction on n, which is the same thing as classifying the amalgamation classes of finite n-tournaments, then we can look forward to a quite smooth inductive analysis along these lines. We have to return now to the basis for the induction. In reality the basis for the induction is the case $n = 3$, because this is the first case in which no new phenomena are encountered within the classification. So for the present section we take $n = 3$, but rather than carrying through the classification in this case we just derive the concrete lemmas needed, and subsequently treat all cases together.

LEMMA 3.8. *Let* $\mathbb{S}$ *be a finite $\mathbb{T}$-restricted 3-tournament with all components linear. Assume that S_1 is a singleton and that S_2 realizes a unique 1-type over S_3. Then* $\mathbb{S}$ *embeds in* $\mathbb{T}$.

PROOF. We proceed by induction on $|S_3|$. If $|S_3| = 1$ this is covered by Lemma 3.6. If $|S_3| > 1$ let c be the first element of S_3, and $L = S_3 - \{c\}$. The proof is completed by applying induction in a derived tournament $\mathbb{T}(c)$ over c. It is only necessary to check that $\mathbb{S}-\{c\}$ is $\mathbb{T}(c)$-restricted. All of the canonical data for $\mathbb{S}-\{c\}$ correspond in $\mathbb{S}$ to 3-tournaments which are either contained in a 2-restriction of $\mathbb{S}$, or to which Lemma 3.6 applies. □

LEMMA 3.9. *Let* $\mathbb{S}$ *be a finite polarized $\mathbb{T}$-restricted 3-tournament with S_1 a singleton, S_2 of order 2, and S_3 linear. Then* $\mathbb{S}$ *embeds in* $\mathbb{T}$.

PROOF. If S_2 realizes a unique type over the rest of $\mathbb{S}$ this is quite easy. Also, if S_2 realizes a source and sink over S_1, and we are dealing with a situation where this imposes a restriction on the orientation of the edge in S_2, then this reduces at once to the case of Lemma 3.6.

Let $S_1 = \{a\}$. In the cases remaining, we perform the following amalgamation:

a_2	y_2	L_1
│	◉	⇑
▼	◉	
a_1	y_1	L_2

arranged so that $(\{a_i\}, \{y_1 y_2\}, L_i)$ must become isomorphic to $\mathbb{S}$ for $i = 1$ or 2.

The factors are of the form:

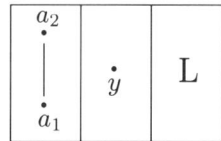

These are $\mathbb{T}$-restricted since either sinks and sources do not occur in (S_1, S_2) or they are of no effect on triangles. In addition we may take it that a_1, a_2 realize a single type over L. We are now in the situation of the previous lemma. □

LEMMA 3.10. *Let $\mathbb{S}$ be a polarized $\mathbb{T}$-restricted 3-tournament with all components linear, and with S_1 a singleton. Then $\mathbb{S}$ embeds in $\mathbb{T}$.*

PROOF. The case $|S_2| \leq 2$ has been treated above. If (S_1, S_2) is also polarized then Lachlan's Ramsey argument applies. The effect of this is that we may take S_2 to be a singleton, at the cost of taking S_1 to be linear and S_3 somewhat longer; and S_1 may be taken to realize a unique type over S_3. This then allows us to proceed by induction on the length of $|S_3|$, passing to derived tournaments, as before.

For the general case we proceed by induction on $|S_2|$. We may suppose that $|S_2| \geq 3$ and that (S_1, S_2) realizes both a sink and a source. Then since S_1 is a singleton we easily form an amalgamation diagram involving the determination of the orientation of an edge in S_2, where the solution is unique. Indeed, in most cases the edge must go from the source to the sink. The exception arises when $T_2 \simeq S(2)$, but since $|S_2| \geq 3$ there is still a suitable forced amalgamation. □

3.7. Some special cases

LEMMA 3.11. *Let $\mathbb{T}^- = (T_1, \ldots, T_{n-1})$, and suppose that $\mathbb{T}^-$ is uniquely determined by its canonical data. Let $\tau : \{1, \ldots, n-1\} \to \{\pm 1\}$. Then $\mathcal{A}(\tau)$ generates $\mathbb{T}^-$, that is, $\mathcal{A}(\tau)$ consists of the finite $(n-1)$-tournaments which embed into $\mathbb{T}^-$.*

PROOF. By the Fraïssé correspondence between amalgamation classes and homogeneous structures, we are merely asserting that the canonical data for $\mathbb{T}^-$ are all in $\mathcal{A}(\tau)$. This may be expressed more explicitly in terms of the following three assertions:

(A): If $\mathbb{A}$ is a transversal of $\mathbb{T}^-$ and $\mathbb{S} = (\mathbb{A}, L)$ is a finite linear extension of $\mathbb{A}$ to a $\mathbb{T}$-restricted and τ-polarized n-tournament, then $\mathbb{S}$ embeds in $\mathbb{T}$.

(B): If A is a finite subtournament of a component of $\mathbb{T}^-$ and (A, L) is a finite linear extension of A to a $\mathbb{T}$-restricted and τ-polarized 3-tournament, then $\mathbb{S}$ embeds in $\mathbb{T}$.

(C): If $\mathbb{A}$ has two components, one a singleton and the other linear, and $\mathbb{S} = (\mathbb{A}, L)$ is a linear extension of $\mathbb{A}$ to a finite $\mathbb{T}$-restricted and τ-polarized 3-tournament, then $\mathbb{S}$ embeds in $\mathbb{T}$.

Now (A) is given by Lemma 3.6. In (B) we are dealing with a 2-restriction of $\mathbb{T}$, so there is no problem. In the third case we have a 3-tournament of the form $\mathbb{S} = (\{x\}, L_2, L_3)$ with L_2, L_3 linear and with $\mathbb{S}$ $\mathbb{T}$-restricted and τ-polarized. This case was treated in the preceding section. □

LEMMA 3.12. *Let $\mathbb{T}$ be a shuffled n-tournament with components isomorphic with $\mathbb{Q}$. Then $\mathbb{T}$ is characterized by its canonical data.*

PROOF. Perhaps it should be stressed that our global assumption that $\mathbb{T}$ is homogeneous remains in force here. Let $\mathbb{S}$ be finite and $\mathbb{T}$-restricted. We proceed by induction on the number of components, and then on the size of $\mathbb{S}$. Let $\mathbb{T}^-$ be the restriction of $\mathbb{T}$ to the first $n-1$ components. Let $\mathbb{S}^-$ be the restriction of $\mathbb{S}$ to its first $n-1$ components. By induction $\mathbb{S}^-$ embeds in $\mathbb{T}^-$ and hence by the previous lemma $\mathbb{S}^-$ is in $\mathcal{A}(\tau)$ for all $\tau: \{1,\ldots,n-1\} \to \{\pm 1\}$. Hence if $\mathbb{S}$ is polarized we are done.

If $\mathbb{S}$ is not polarized and specifically there is both an (i,n)-source and an (i,n)-sink realized in $\mathbb{S}$, then there is also an element x in $S_i \cup S_n$ realizing both an (i,n)-source and an (i,n)-sink, so there is a straightforward amalgamation with $\mathbb{S}$ as its unique solution, whose factors have smaller size. □

LEMMA 3.13. *Let $\mathbb{T}$ be a shuffled n-tournament whose components are isomorphic to $S(2)$. Then $\mathbb{T}$ is determined by its canonical data.*

PROOF. Up to a variant (reorientation of the edges in some components) $\mathbb{T}$ may be thought of as a single copy of $S(2)$ which has been partitioned into n dense subsets, and has subsequently had the edges between any pair of components colored by some finite set of colors. Our claim is that the final step took place generically; this somewhat metaphysical statement means that any finite n-tournament embedding in the generic n-tournament of this type will embed in $\mathbb{T}$. The geometry of the situation is useful.

As a matter of notation, the types occurring between two components T_i, T_j have been partitioned into two sets L_{ij}, R_{ij}. A more precise way of formulating our opening remarks is that we can suppose that the partition is the same if i, j are switched, and that the structure $\bar{\mathbb{T}}$ in which we replace L_{ij}, R_{ij} by single types l_{ij}, r_{ij} is the one induced on $S(2)$ by a partition into n dense subsets, and the original edge relation.

Let $\mathbb{S}$ be a finite $\mathbb{T}$-restricted tournament. We shall show that $\mathbb{S}$ embeds in $\mathbb{T}$ by induction, first on the number of components n of $\mathbb{T}$, and then on the number m of components of $\mathbb{S}$ with more than one element. We may suppose that $\mathbb{S}$ is not a transversal, and fix i so that $|S_i| > 1$. Let $\mathbb{S}^-$ be the restriction of $\mathbb{S}$ to the other components. For $a \in S_i$ let $p(a)$ be $\mathrm{tp}(a/\mathbb{S}^-)$ and let $\bar{p}(a)$ be the type of a over $\mathbb{S}^-$ in $\bar{\mathbb{S}}$, that is the conjunction of all true conditions of the form:

"$\mathrm{tp}(xb) \in L_{ij}$" or "$\mathrm{tp}(xb) \in R_{ij}$"

which hold for $x = a$ and for $b \in \mathbb{S}^-$, together with the condition that x lies in the i-th component. So $\bar{p}(a)$ is the expected type of a in $\bar{\mathbb{T}}$, once our embedding is found.

By induction on m, $\mathbb{S}^- \cup \{a\}$ embeds in $\mathbb{T}$. Then relative to a fixed embedding ι_a, $\bar{p}(a)$ is the definition of an interval in T_i. We may assume that all of the maps ι_a ($a \in S_i$) agree on $\mathbb{S}^-$, by homogeneity. Now there is an isomorphic embedding ι of S_i into T_i such that $\iota(a)$ satisfies $\bar{p}(a)$ for each a. Any perturbation of this embedding by a small amount has the same property. So it suffices to prove for each a:

The set defined by $p(a)$ is dense in the set defined by $\bar{p}(a)$.

If this is false then there is an $(\mathbb{S}^-)$-definable cut in S_i distinct from the ones associated with intervals of the type of $\bar{p}(a)$. We have canonical shuffling isomorphisms between the dedekind completions of the components of $\mathbb{T}$, so our cut can be transported into T_j for any $j \neq i$ via such an isomorphism. This means that the situation is visible in $\mathbb{T}^-$. By induction on n, $\mathbb{T}^-$ is the canonical shuffled tournament with a certain set of cross types, so we have a contradiction. □

3.8. The general case

We now treat the case of the general homogeneous n-tournament $\mathbb{T}$. We have previously excluded diagonal and finite components. We can quickly exclude components of type $\mathbb{Q}, S(2)$ as well.

LEMMA 3.14. *Let (I, J) be a partition of $\{1, \ldots, n\}$ into two proper subsets such that for $i \in I$, $j \in J$ there are no (i,j)-sources and no (i,j)-sinks. Suppose that every $(n-1)$-restriction of $\mathbb{T}$ is characterized by its canonical data. Then $\mathbb{T}$ is also characterized by its canonical data.*

PROOF. We let $\mathcal{A}(I)$ be the set of finite I-tournaments (that is, with components indexed by I) $\mathbb{S}$ embedding in $\mathbb{T} \upharpoonright I$ such that for every J-tournament $\mathbb{A} \leq \mathbb{T} \upharpoonright J$:

Any $\mathbb{T}$-restricted extension of $\mathbb{S}$ by $\mathbb{A}$ embeds in $\mathbb{T}$.

The absence of sources and sinks makes this an amalgamation class.

As $\mathbb{T} \upharpoonright I$ is determined by its canonical data, it suffices to show that these I-tournaments lie in $\mathcal{A}(I)$. We may suppose that $|I| > 1$ (replacing I by J if need be).

So fix $\mathbb{S} \leq \mathbb{T} \upharpoonright I$ one of the tournaments belonging to the canonical data for $\mathbb{T} \upharpoonright I$, let $\mathbb{A} \leq \mathbb{T} \upharpoonright J$ be finite, and let $(\mathbb{S}, \mathbb{A})$ denote a $\mathbb{T}$-restricted extension of $\mathbb{S}$ by $\mathbb{A}$. We shall show by induction on n and $|\mathbb{S}|$ that $(\mathbb{S}, \mathbb{A})$ embeds in $\mathbb{T}$. If $\mathbb{S}$ is τ-polarized at the ith component, where $\tau : I - \{i\} \to \{\pm 1\}$, then $(\mathbb{S}, \mathbb{A})$ is also polarized at the ith component, with respect to any extension of τ to $\{1, \ldots, \hat{i}, \ldots, n\}$. In this case by Lemma 3.11 $(\mathbb{S}, \mathbb{A})$ embeds in $\mathbb{T}$. This holds in particular if $\mathbb{S}$ is a transversal or is contained in a single component of $\mathbb{T}$. Therefore we may now suppose that $\mathbb{S}$ has only one component $S_{i'}$ other than S_i and that this component is linear of order at most 3.

If $\mathbb{S}$ is not polarized at i, then in most cases there is an amalgamation with a unique solution producing $(\mathbb{S}, \mathbb{A})$ from factors whose I-restrictions are smaller. The exceptional case is that in which $T_i \simeq S(2)$. If $|S_{i'}| = 3$ there is still such an amalgamation. If $|S_{i'}| = 2$ we write $S_i = \{a\}$, $S_{i'} = \{y_1, y_2\}$, and we may suppose that $\text{tp}(ay_1) \in L_{ii'}$, $\text{tp}(ay_2) \in R_{ii'}$. We use an amalgamation of the following sort:

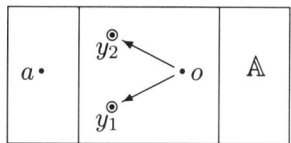

Here o is an auxiliary point chosen so that $\text{tp}(ao) \in L_{ii'}$ and the whole picture is $\mathbb{T}$-restricted. The critical factor here is the one omitting y_1, which is built by amalgamating aoy_2 and $ay_2\mathbb{A}$ over ay_2. The factor omitting y_2 is then $\mathbb{T}$-restricted and is also polarized at i, hence covered by Lemma 3.11. $\square$

Finally we prove that *any homogeneous n-tournament is uniquely characterized by the associated canonical data: components, transversals, and additional data pinning down the 2-restrictions.*

PROOF. We proceed by induction on n. We may assume that $n \geq 3$, $\mathbb{T}$ has no finite components or diagonal 2-restrictions. The relation on $\{1, \ldots, n\}$ defined by "(T_i, T_j) is shuffled" is an equivalence relation. If it has only one class, then we have already shown that $\mathbb{T}$ is characterized by its canonical data. If there is more than one class, and for some i T_i is isomorphic to $\mathbb{Q}$ or $S(2)$, then letting I be the class of i, and J the remaining indices, we arrive at the hypotheses of the preceding lemma.

Accordingly we may now assume that every component of $\mathbb{T}$ is isomorphic with T^∞. With each function $\tau : \{1,\ldots,n-1\} \to \{\pm 1\}$ we associate the class $\mathcal{B}(\tau)$ of all finite n-subtournaments $\mathbb{S}$ of $\mathbb{T}$ such that any $\mathbb{T}$-restricted linear extension $\mathbb{S}^+ = \mathbb{S} \cup L$ of $\mathbb{S}$ taken so that $(S_1,\ldots,S_{n-1},L)$ is τ-polarized will embed in $\mathbb{T}$. Like $\mathcal{A}(\tau)$ (where we required $S_n = \emptyset$), this is an amalgamation class. We claim the canonical data for $\mathbb{T}$ all lie in $\mathcal{B}(\tau)$. Now in all cases the n-tournament $\mathbb{S}^+$ will have a component S_i with at most one element. If $\mathbb{S}^+$ is polarized at i then Lemma 3.11 applies, and otherwise there is an amalgamation whose unique solution is $\mathbb{S}^+$ and whose factors are of smaller order.

Thus every n-tournament occurring in the canonical data for $\mathbb{T}$ is in the class $\mathcal{B}(\tau)$ for all τ. We shall now check that every $\mathbb{T}$-restricted tournament belongs to every amalgamation class containing the canonical data for $\mathbb{T}$, and hence embeds in $\mathbb{T}$. The argument is by induction on $|S_n|$, and then on the order of $\mathbb{S}$. If S_n is empty we apply the induction hypothesis on n. Otherwise we let $\mathbb{S}^-$ be derived from $\mathbb{S}$ by deleting one element $x \in S_n$. There are two cases. If $(S_1,\ldots,S_{n-1},x)$ is τ-polarized for some τ, then by our induction hypothesis $\mathbb{S}^-$ belongs to every amalgamation class containing the canonical data for $\mathbb{T}$, hence also to $\mathcal{B}(\tau)$. Hence $\mathbb{S} = \mathbb{S}^- \cup \{x\}$, as a suitably polarized linear extension of $\mathbb{S}^-$, must embed in $\mathbb{T}$. Alternatively, if $(S_1,\ldots,S_{n-1},x)$ is not polarized, we argue by induction on the order of $\mathbb{S}$, using an amalgamation with a unique solution. In the present case we are not troubled by ambiguous behavior of sinks and sources, as no component is of the form $S(2)$. □

CHAPTER 4

Homogeneous symmetric graphs

4.1. The theorem of Lachlan and Woodrow

In the present chapter we shall give a new proof of a theorem of Lachlan and Woodrow which classifies the infinite homogeneous symmetric graphs. Though we use some ideas which made their first appearance in [61], we avoid one aspect of their argument which does not seem to be adaptable to the context of directed graphs. The proof given here is more complex than the one given in [61], but it generalizes to cover that portion of the analysis of the asymmetric case which is given in Chapters 7 and 8 below, and it illustrates the main ideas of Chapters 7 and 8 in a technically simpler setting. However nothing in later chapters depends on the material given here.

There are other issues arising in the directed case that have no parallel here. These other issues are dealt with in Chapters 5 and 6; they are much like those occurring in the classification of the homogeneous tournaments.

In the remainder of the present chapter the term "graph" will refer exclusively to symmetric graphs with no loops or multiple edges. We shall use the following notation:

G^c	the complement of the graph G
$K(n)$	the complete graph on n vertices
I_n	$K(n)^c$
$G[H]$	the composition of G with H (H replaces the points of G)
P_3	A path of length 2 (order 3), i.e. $(K(1)+K(2))^c$.
$-, \perp$	the two nontrivial 2-types in a homogeneous graph (linked or unlinked)
a'	The set of neighbors of a in a given graph
$a^\perp$	The set of nonneighbors of a in a given graph (except a)

THEOREM 4.1. *[Lachlan, Woodrow] Let G be an infinite homogeneous graph. Then either G or G^c has one of the following forms:*

1. *$I_m[K(n)]$ with $\max(m,n) = \infty$;*
2. *Generic omitting $K(n+1)$;*
3. *Generic.*

In the second or third cases, the meaning is that the finite graphs occurring as subgraphs of G are: exactly those omitting $K(n+1)$ (in the second case) or else all finite graphs whatsoever (in the third case). The finite homogeneous graphs have also been classified [36, 83, 39], but we shall not discuss this here.

For the proof it is more convenient to express the result in terms of amalgamation classes. Let $\mathcal{A}(n)$ be the following class of finite graphs: $\{(K(1)+K(2)), P_3\} \cup \{I_k : \text{all } k\} \cup \{K(n)\}$. Theorem 4.1 may be expressed as follows:

THEOREM 4.2. *Let A be a finite graph which does not embed $K(n+1)$. Then $\mathcal{A}(n) \Longrightarrow A$.*

To deduce Theorem 4.1 from Theorem 4.2 one argues as follows. Let G be a homogeneous graph. If G omits P_3 then it is easily seen to be of the form $I_m[K(n)]$ for some $m, n \le \infty$. The same applies to G^c if G omits $K(1) + K(2)$. (Note also that G is homogeneous iff G^c is.) Now suppose that G is infinite, and embeds $K(1) + K(2)$ and P_3. By Ramsey's theorem G embeds $K(\infty)$ or I_∞, and passing to the complement if necessary, we may suppose the latter occurs. Now applying Theorem 4.2 to the amalgamation class $\mathcal{A}$ of finite graphs embeddable in G, we see that a finite graph A embeds in G if and only if every complete subgraph of A embeds in G. Thus G falls under the second or third case of Theorem 4.1.

The remainder of this chapter is devoted to a proof of Theorem 4.2. The proof proceeds by induction on n, and to keep the induction going we prove two other statements at the same time. These auxiliary theorems concern homogeneous 2-graphs. A 2-graph will be taken to be a structure with two components, such that each component is a graph, and with a finite but arbitrary set of binary relations available as cross types between the components. The important homogeneous 2-graphs will be those derived from ordinary graphs by selecting two 1-types over some finite set; these have at most two cross types.

The following terminology is very useful.

NOTATION 2. 1. *When speaking of 2-graphs, a 1-type over the graph A will mean a 2-graph $\mathbb{A}$ for which A_1 is a singleton and $A_2 \sim A$ (or $A_2 = A$ if the context requires it). Typically there will be a specific set of cross types allowed, generally consisting of all those present in some 2-graph currently under consideration.*

2. *A 2-graph $\mathbb{H}$ will be said to be* ample *if it embeds all 1-types over I_n for all n (with the stated convention on cross types), and if, in addition, $K(1) + K(2)$ and P_3 embed in H_2.*

3. *If $\mathbb{A}$ is a 1-type over A and $B \subseteq A$, then the* restriction *of $\mathbb{A}$ to B is the 1-type (A_1, B) over B induced by $\mathbb{A}$.*

4. *If $\mathbb{H}$ is a 2-graph and $\mathbb{A}$ is a 1-type over A, $\mathbb{A}$ will be said to be $\mathbb{H}$-constrained if for every complete subgraph K of A, the restriction of $\mathbb{A}$ to K embeds in $\mathbb{H}$.*

We can now state our two auxiliary results, which will be proved by simultaneous induction on n.

THEOREM 4.3. *For each n, if $\mathbb{H}$ is an ample homogeneous 2-graph and $\mathbb{A}$ is an $\mathbb{H}$-constrained 1-type over $\oplus_i B_i$, where each $B_i \in \mathcal{A}(n)$, then $\mathbb{A}$ embeds in $\mathbb{H}$.*

THEOREM 4.4. *For each n, if $\mathbb{H}$ is an ample homogeneous 2-graph and $\mathbb{A}$ is an $\mathbb{H}$-constrained 1-type over a subgraph A of H_2 which omits $K(n+1)$, then $\mathbb{A}$ embeds in $\mathbb{H}$.*

At the n-th stage of our inductive argument, Theorems 4.2, 4.3 and 4.4 will be proved for n.

One point that deserves considerable attention concerns the application of Ramsey's theorem. In the previous chapters, whenever we have applied Ramsey's theorem to extract a well-behaved structure from a somewhat larger structure, we have always known that we would wind up with a long linear tournament, and indeed we have banked on it. When applied to an infinite graph, Ramsey's theorem delivers either $K(\infty)$ or I_∞. This ambiguity has an impact, in the first place on our notation, and in the second place, on our strategy. The relevant notation will be established when the issue arises, at the beginning of §4.4 below.

4.2. The main ingredients

We shall now give a more detailed outline of our proof of the classification theorem for homogeneous symmetric graphs. We formulate four quite concrete propositions and three more general theorems to be proved by induction. The latter constitute the main steps in the derivation of our three main theorems. Typically the most concrete results depend on very explicit amalgamation arguments, and then the more general results follow on formal grounds.

We make a general observation that will be applied tacitly later on. Let $\mathcal{A}$ be a set of finite graphs, B a finite graph, and suppose $\mathcal{A} \implies B + A$ for all $A \in \mathcal{A}$. Then $\mathcal{A} \implies B + A$ for all A such that $\mathcal{A} \implies A$. One way to see this is to consider a homogeneous graph H embedding all elements of $\mathcal{A}$, to fix a copy of B in H, and to consider $B^\perp$. In a similar vein, if $\mathcal{A} \implies A_1 + A_2$ when $A_1, A_2 \in \mathcal{A}$, then also $\mathcal{A} \implies A_1 + A_2$ when A_1, A_2 are consequences of $\mathcal{A}$. (The previous result may be applied twice.)

Our next result concerns the following construction. If A is a finite graph, let H^+ be the graph with vertex set $V(H) \times \{0, 1\}$ in which the induced graph on $V(H) \times \{0\}$ is a copy of H, there is an edge linking each vertex $(a, 0)$ to the vertex $(a, 1)$, and there are no other edges.

LEMMA 4.5. *Let $\mathcal{A}$ be a set of finite graphs such that for $A \in \mathcal{A}$ we have $\mathcal{A} \implies A^+$ and $\mathcal{A} \implies A + I_1$. Then $\mathcal{A} \implies A^+$ whenever $\mathcal{A} \implies A$.*

PROOF. This depends on the fact that if $\mathcal{A} \implies A$ then there is a finite tree of amalgamation diagrams, with the branching at each node corresponding to all the possible completions of the diagram at the given node, and with the factors in each diagram coming from $\mathcal{A}$ or from the assumed results of the diagrams along earlier nodes, so that along each maximal branch of the tree, A is contained in some configuration associated with that branch. Compare [**61**].

In addition we require the following observation. Let A_1, A_2 be extensions of A_0, and consider the possible amalgams of $A_1^+ + (A_2^+ - A_2), A_2^+ + (A_1^+ - A_1)$ over $A_0 + (A_1^+ - A_1) + (A_2^+ - A_2)$. This has factors of the forms $I_n + A_i^+$. Let $\mathcal{A} + I_n = \{A + I_n : A \in \mathcal{A}\}$ and similarly $\mathcal{A}^+ = \{A^+ : A \in \mathcal{A}\}$. Then from a tree of amalgamation diagrams witnessing $\mathcal{A} \implies A$ we derive a tree witnessing $I_n + \mathcal{A}^+ \implies A^+$ for some n, and by hypothesis every graph in $I_n + \mathcal{A}^+$ is a consequence of $\mathcal{A}$. □

Now we can formulate the main ingredients of our proof explicitly.

PROPOSITION 10. *If $\mathcal{A}(0) \implies A$ then $\mathcal{A}(0) \implies K(2) + A$.*

PROPOSITION 11. *If $\mathcal{A}(0) \implies A$ then $\mathcal{A}(0) \implies A^+$.*

PROPOSITION 12. *If $\mathbb{H}$ is an ample homogeneous 2-graph and $\mathbb{A}$ is an $\mathbb{H}$-constrained 1-type over $2 \cdot K(2)$, then $\mathbb{A}$ embeds in $\mathbb{H}$.*

PROPOSITION 13. *If $\mathbb{H}$ is an ample homogeneous 2-graph and $\mathbb{A}$ is an $\mathbb{H}$-constrained 1-type over P_3, then $\mathbb{A}$ embeds in $\mathbb{H}$.*

THEOREM 4.6. $\mathcal{A}(n) \implies \mathcal{A}(n) + \mathcal{A}(n)$

THEOREM 4.7. *Let $\mathbb{H}$ be an ample homogeneous 2-graph with $K(n)$ embedding in H_2, and let $\mathbb{A}$ be an $\mathbb{H}$-constrained 1-type over $2 \cdot [K(n)]$. Then $\mathbb{A}$ embeds in $\mathbb{H}$.*

THEOREM 4.8. *Let $K \simeq K(n)$ and let $A = \{a\} \cup K$ with a linked to a unique vertex in K. Then $\mathcal{A}(n) \implies A$.*

THEOREM 4.9. *If A is the disjoint union of graphs in $\mathcal{A}(n)$, A^* is an extension of A by a single point, and A^* omits $K(n+1)$, then every homogeneous graph embedding the elements of $\mathcal{A}(n)$ must also contain A^*.*

4.3. Structure of the proof

We now give a precise formulation of the steps of the proof.

1. Prove the four propositions, 10–13, in order.
2. Assume Theorem 4.2 holds for graphs A omitting $K(n)$. Prove Theorem 4.8 for n.
3. Assume Theorem 4.4 for $n-1$ and Theorem 4.8 for n. Prove Theorem 4.6 for n.
4. Assume Theorems 4.6 and 4.8 for n, and assume Theorem 4.4 holds for graphs A omitting $K(n)$. Prove Theorem 4.7 for n.
5. Assume Proposition 13 and Theorems 4.6 and 4.7 for n, where $n \geq 2$. Prove Theorem 4.3 for n.
6. Assume Propositions 11 and 13, and assume Theorem 4.4 for graphs A omitting $K(n)$ and Theorem 4.8 for $m < n$, as well as Theorem 4.3 for n. Derive Theorems 4.8 and 4.9 for n; in the proof of Theorem 4.9, Theorem 4.8 is used.
7. Assume Theorem 4.9 for n. Prove Theorem 4.2 for n.
8. Assume Theorems 4.2 and 4.3 for n. Prove Theorem 4.4 for graphs A omtting $K(n+1)$.

We shall carry out these seven steps in the order: 7, 5, 8; 1; 6, 2, 3, 4; that is: we show how the overall induction operates, we dispose of the four concrete propositions, and then we enter into the details.

4.4. Steps 7, 5, 8. Proof of the Main Theorems

We shall carry out steps 7, 5, and 8 of our outline above. With some abuse of language we may say that we are giving the derivation of Theorems 4.2, 4.3, and 4.4 from our four propositions and three more technical theorems. This is the gist of the matter, though this formulation ignores the form of the induction. Matters have been organized so that this part of the proof is largely a formal exercise.

Step 7. Assume Theorem 4.9 for n. Prove Theorem 4.2 for n.

Our hypothesis is:

> If A is the disjoint sum of graphs in $\mathcal{A}(n)$, A^* is an extension of A by a single point, and A^* omits $K(n+1)$, then every homogeneous graph embedding the elements of $\mathcal{A}(n)$ must also contain A^*.

The desired conclusion is:

Let A be a finite graph not embedding $K(n+1)$. Then $\mathcal{A}(n) \implies A$.

Here Ramsey arguments will play a role.

DEFINITION 4.10. Let r be a nontrivial 2-type. A *ramsey graph* of type r is a graph in which any two distinct elements realize the type r; in other words, a clique or an independent set of vertices.

DEFINITION 4.11. With n fixed, let $\mathcal{A}$ be an amalgamation class containing $\mathcal{A}(n)$, and let r be a 2-type. We define $\mathcal{A}^r$ to be the set of all $A \in \mathcal{A}$ satisfying:

(r): if $R \cup A$ is any extension of A by a finite r-ramsey graph R such that $\{x\} \cup A$ omits $K(n+1)$ for all x in R, then $R \cup A \in \mathcal{A}$.

In the notation the dependence of $\mathcal{A}^r$ on n is not shown. In practice one normally has a definite value of n.

The following is immediate and useful.

LEMMA 4.12. *If n is fixed, $\mathcal{A}$ is an amalgamation class containing $\mathcal{A}(n)$, and r is a nontrivial 2-type, then $\mathcal{A}^r$ is an amalgamation class.*

The proof of the next theorem uses Theorem 4.9 and Lachlan's Ramsey argument, and is immediate modulo those ingredients:

LEMMA 4.13. *If $\mathcal{A}$ is an amalgamation class containing $\mathcal{A}(n)$ then there is a 2-type r for which $\mathcal{A}^r$ contains $\mathcal{A}(n)$.*

PROOF. If we suppose the contrary, then for each nontrivial 2-type r we have a graph $A_r \in \mathcal{A}(n)$ and a finite r-Ramsey extension $B_r = R_r \cup A_r$ of A_r such that $\{x\} \cup A_r$ omits $K(n+1)$ for all $x \in R_r$, while B_r is not in $\mathcal{A}$. Let $A = \oplus_r A_r$ (the disjoint union of two such graphs). We apply Proposition 4.9 to this graph: any extension of $k \cdot A$ by a single point which omits $K(n+1)$ will lie in $\mathcal{A}$, and hence by Lachlan's Ramsey argument, for k large this forces $R_r \cup A_r$ to be in $\mathcal{A}$ for some r, a contradiction. The new circumstance, that we cannot predict which type r will be selected, has no effect on the course of the argument. □

We may now carry out the formal portion of the proof of Theorem 4.2.

PROOF. We prove by induction on k:

> If $\mathcal{A}$ is an amalgamation class containing $\mathcal{A}(n)$, and A is a graph with k vertices omitting $K(n+1)$, then $A \in \mathcal{A}$.

For $k = 1$ this is clear. For $k > 1$, with $\mathcal{A}$ and A fixed choose $x \in A$ and r a 2-type so that $\mathcal{A}(n) \subseteq \mathcal{A}^r$. Let $A^* = A - \{x\}$. By induction on k, $A^* \in \mathcal{A}^r$, and setting $R_r = \{x\}$ we find that $A = R_r \cup A^*$ belongs to $\mathcal{A}$, as claimed. □

This completes Step 7.

Step 5. With $n \geq 2$ fixed, assume Propositions 10 and 13, and assume Theorems 4.6 and 4.7 for n. Prove Theorem 4.3 for n.

Actually, for later use we shall prove something slightly more precise. Assume Propositions 10 and 13, and Theorem 4.6 for n. Let $\mathbb{H}$ be a homogeneous ample 2-graph, and suppose that Theorem 4.7 applies to every 2-graph $\mathbb{H}^*$ of the form

($^pA, A^\perp$) with $A \subseteq H_2$, and p a 1-type over A realized in H_1. Then we show that Theorem 4.3 applies to $\mathbb{H}$.

So in keeping with this more precise formulation, $\mathbb{H}$ is fixed in this subsection.

LEMMA 4.14. *Let $A \leq H_2$ with $A \simeq P_3$ or $K(n)$, and let p be a 1-type over A which is realized in H_1. Then $\mathbb{H}(A) := (^pA, A^\perp)$ is ample, and has the same cross types as $\mathbb{H}$.*

PROOF. By Theorem 4.6, $A^\perp$ contains $K(1) + K(2)$ and P_3, so it suffices to check that every type over I_k whose cross types occur in $\mathbb{H}$ is realized in $\mathbb{H}(A)$, for all k. We show by induction on k that every $\mathbb{H}$-constrained 1-type over $A + I_k$ embeds in $\mathbb{H}$. For $k = 0$ this is trivial if $A \simeq K(n)$, and it is Proposition 13 if $A \simeq P_3$.

For $k > 1$ let $A_1 = \{x\}$, fix $y \in I_k \subseteq A_2$, and let $p = \text{tp}(xy)$. Identify y with an element of H_2 and let $\mathbb{H}^* = (^py, y^\perp)$. Clearly $\mathbb{H}^*$ is ample, and by induction we need only check that $\mathbb{A} - \{y\}$ is $\mathbb{H}^*$-constrained. This follows from Theorem 4.7. $\square$

We shall prove Theorem 4.3 for $\mathbb{H}$ in the following form, by induction on m:

If $A = \oplus_i B_i$ is the disjoint union of m graphs B_i, all of the form $K(n)$ or P_3, then every $\mathbb{H}$-constrained 1-type $\mathbb{A}$ over A embeds in $\mathbb{H}$.

In the proof we may replace $\mathbb{H}$ by a 2-graph $\mathbb{H}^* = (^pA, A^\perp)$ as the same hypotheses continue to apply.

PROOF. For $m = 1$ the claim is vacuous when $A = K(n)$, and for $A = P_3$ our claim is stated as Proposition 13, which we are presently assuming.

For $m > 1$ let p be the restriction of $\mathbb{A}$ to B_1 and let $\mathbb{H}^* = (^pB_1, B_1^\perp)$, taking $B_1 \subseteq H_2$. By the previous lemma $\mathbb{H}^*$ is ample and has the same 1-types over elements of $K(n)$ and P_3 as $\mathbb{H}$ does. It suffices to check that $\mathbb{A} - B_1$ is $\mathbb{H}^*$-constrained, as an application of the inductive hypothesis then embeds $\mathbb{A} - B_1$ into $\mathbb{H}^*$, and hence embeds $\mathbb{A}$ into $\mathbb{H}$. This means that we need only treat the case $m = 2$, with B_2 a complete graph. However in this case we may interchange B_1 and B_2 and repeat the argument, reducing to the case covered by Theorem 4.7. $\square$

Step 8. Assuming Theorem 4.2 and Theorem 4.3, prove Theorem 4.4.

PROOF. We fix an ample homogeneous 2-graph $\mathbb{H}$. We want to prove that every $\mathbb{H}$-constrained 1-type over a graph omitting $K(n+1)$ will embed in $\mathbb{H}$. As in the proof of Lemma 4.13, we may deduce from Theorem 4.3 that there is a 2-type r realized in H_1 such that for every finite $\mathbb{H}$-constrained configuration (R, A) with R r-ramsey and $A = \oplus B_i$ constructed from $B_i \in \mathcal{A}(n)$, the 2-graph (R, A) embeds in $\mathbb{H}$. If we let $\mathcal{A}$ be the class of all finite graphs A which satisfy this extension property, then $\mathcal{A}$ is an amalgamation class containing $\mathcal{A}(n)$. By Theorem 4.2 it follows that $\mathcal{A}$ contains all graphs omitting $K(n+1)$, which implies our claim. $\square$

4.5. Step 1, Proposition 10: adding $K(2)$

We shall show in this section that if $\mathcal{A}(0) \implies A$, then $\mathcal{A}(0) \implies A + K(2)$.

LEMMA 4.15. *Let H be a primitive homogeneous graph embedding I_4 and $K(2)$. Then H embeds $I_2 + K(2)$.*

PROOF. By primitivity H embeds $I_1 + K(2)$. Suppose that H omits $I_2 + K(2)$, and fix $a \in H$. Then $\perp$ is an equivalence relation on $a^\perp$ with classes of size at least 3. For $x \in a'$, if C is a $\perp$-class of $a^\perp$ we have $|x^\perp \cap C| \leq 1$ as H omits $I_2 + K(2)$. By homogeneity it follows that $x^\perp$ meets each $\perp$-class in $a^\perp$ in a unique point. In particular $2 \cdot K(2) \leq H$. For any $x, y \in a'$, and any $\perp$-class C of $a^\perp$, $x' \cap y' \cap C$ is nonempty; since it is possible to choose x, y, C so that $x^\perp \cap C \neq y^\perp \cap C$, it follows that for any distinct $x, y \in a'$, $x^\perp \cap a^\perp$ and $y^\perp \cap a^\perp$ are disjoint. However by homogeneity any two linked elements of $a^\perp$ occur in $x^\perp$ for some $x \in a'$, a contradiction. □

LEMMA 4.16. *If* $\mathcal{A}(0) \implies A$ *then* $\mathcal{A}(0) \implies I_1 + A$.

PROOF. By the previous lemma $\mathcal{A}(0) \implies I_2 + K(2)$. All we need to show is that $\mathcal{A}(0) \implies I_1 + P_3$. Suppose H is a homogeneous graph embedding P_3, $K(1) + K(2)$, and I_∞, but H omits $I_1 + P_3$. Then for $a \in H$, $a^\perp \simeq I_\infty[K(m)]$ for some m, $2 \leq m \leq \infty$. Thus we may speak of the components of $a^\perp$. In addition, as H omits $I_1 + P_3$ one sees easily that for $a \perp b$, the components of $a^\perp$ and $b^\perp$ coincide, except for the components containing a and b themselves. As G^c is connected, for any two vertices a, b, we see that $a^\perp$ and $b^\perp$ differ by at most finitely many components. In particular if a, b, c form a path P_3, then $a^\perp \cap b^\perp \cap c^\perp$ is nonempty, a contradiction. □

LEMMA 4.17. $\mathcal{A}(0) \implies 2 \cdot K(2)$.

PROOF. We use the following amalgamations. If the first fails, the second one succeeds.

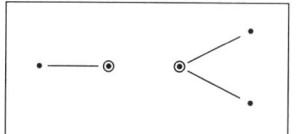 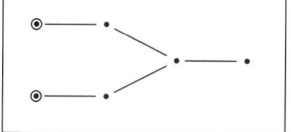

□

LEMMA 4.18. $\mathcal{A}(0) \implies K(2) + P_3$

PROOF. Let $\mathcal{A}$ be an amalgamation class containing $\mathcal{A}(0)$ and omitting $K(2) + P_3$. Then easily $\mathcal{A}$ contains:

(1) 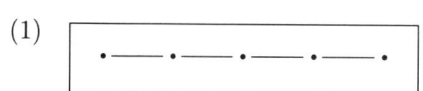 (2)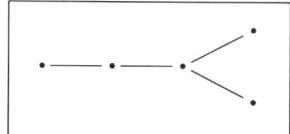

We shall now show that $\mathcal{A}$ also contains:

(3)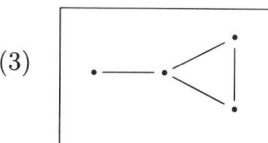

If $\mathcal{A}$ omits $K(2) + P_3$ then we can easily form the factors of the amalgamation diagram:

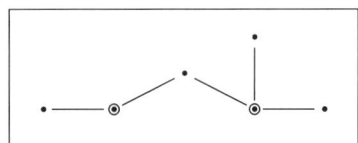

and this produces $K(2) + P_3$ or (3). Thus we may suppose that (3) belongs to $\mathcal{A}$.

Now we let $\mathcal{A}^* = \{A : A + I_1 \in \mathcal{A}\}$, which is also an amalgamation class containing $\mathcal{A}(0)$ and omitting $K(2) + P_3$. So (1), (2), (3) belong to $\mathcal{A}^*$. If $\mathcal{A}$ also contains $K(2) + K(3)$ then we can form:

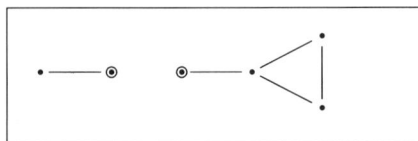

and this produces $K(2) + P_3$. If $\mathcal{A}$ does not contain $K(2) + K(3)$ then we easily get the factors for:

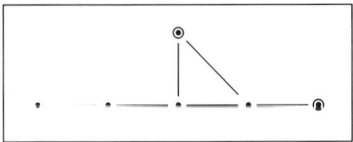

and this produces $K(2) + P_3$ or $K(2) + K(3)$. □

Comment We used Lemma 4.16 in the preceding proof. For this reason we needed to look at the auxiliary class $\mathcal{A}^*$. The effect of this was to treat $\mathcal{A}$ as if $\mathcal{A} \supseteq I_1 + \mathcal{A}$; we have no reason as yet to believe this is literally true, but as we saw here it might as well be true. In later arguments we shall compress this line of reasoning, and avoid the explicit introduction of the relevant auxiliary class.

Now Proposition 10 follows from the preceding lemma and Lemma 4.16.

4.6. Step 1, Proposition 11: the operation H^+

We recall that the attachment operation A^+ attaches edges systematically to the vertices of A. Proposition 11 says that the set of consequences of $\mathcal{A}(0)$ is closed under this operation. In our proof we shall make use of Proposition 10.

Let $[I_1, I_3]$ be the graph $(K(1) + K(3))^c$.

LEMMA 4.19. *Any amalgamation class of finite graphs containing $\mathcal{A}(0)$ contains $[I_1, I_3]$.*

PROOF. We use Proposition 10 freely.

Let H be a homogeneous graph embedding the graphs of $\mathcal{A}(0)$ and omitting $[I_1, I_3]$. We deal first with a very special case. If H has a path of length 3 and no path of length 4 we get a contradiction from:

Now we turn to the general case. One easily obtains:

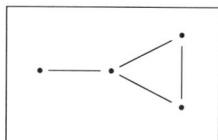

(One sets up two amalgamations, depending on whether H embeds a path of length 3 or not).

Then using Proposition 10 one can form:

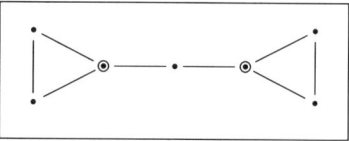

In particular H contains a path of length 3, and hence a path of length 4. No matter how this diagram is completed, one sees that H contains the factors for the following amalgamation:

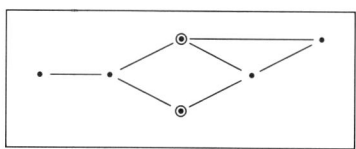

Here an edge must be inserted. Hence we can set up the following final amalgamation:

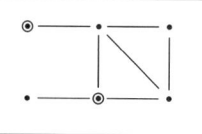

This one produces a copy of $[I_1, I_3]$ in all cases. □

LEMMA 4.20. $\mathcal{A}(0) \implies P_3^+$.

PROOF. We use the following amalgamations:

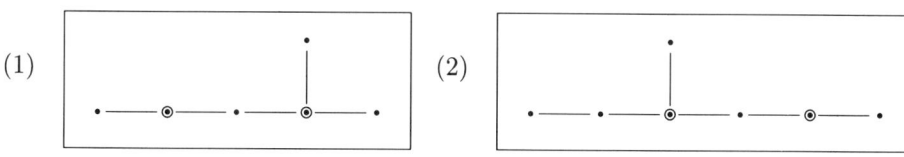

We can set up (1) using the previous Lemma (and Lemma 4.16, which we have been using freely throughout). The result of (1) can be used to set up (2). □

Now we prove Proposition 11.

PROOF. By Lemmas 4.16 and 4.20 we need only show that $\mathcal{A}(0) \implies I_n^+$ and $(K(1)+K(2))^+$. As $I_n^+ = n \cdot K(2)$, and $(K(1)+K(2))^+$ is contained in $K(2)+P_3^+$, the preceding lemma combined with Proposition 10 yields the result. □

4.7. Step 1, Propositions 12 and 13: realization of 1-types

In the present section $\mathbb{H}$ denotes an ample homogeneous 2- graph. We are concerned with the realization of 1-types over $2 \cdot K(2)$ and P_3.

Recall that a 2-graph $\mathcal{A}$ is called $\mathbb{H}$-*constrained* if each of its restrictions to a 1-type over any complete subgraph embeds in $\mathbb{H}$. (Since $\mathbb{H}$ is ample, we need no hypotheses relative to 1-types over I_n.) A 2-type r realized in H_1 will be called a *ramsey type* for the subgraph H of H_2 if every finite $\mathbb{H}$-constrained 2-graph of the form (R, A) with R r-ramsey and $A \subseteq H$ embeds in $\mathbb{H}$.

LEMMA 4.21. *There is a ramsey type over I_∞.*

PROOF. As $\mathbb{H}$ is assumed ample this is a consequence of Lachlan's Ramsey argument. Much the same point was made in the proof of Lemma 4.13. □

LEMMA 4.22. *Every $\mathbb{H}$-constrained 1-type $\mathbb{A}$ over $I_1 + K(2)$ embeds in $\mathbb{H}$.*

PROOF. Let r be a ramsey type over I_∞. Let $R = \{x_1, x_2\}$ be r-ramsey. Let $B_0 = \{a, b\} \simeq I_2$, $B_i = B_0 \cup \{y_i\}$ for $i = 1, 2$ with $B_1 \simeq I_3$, $B_2 \simeq I_1 + K(2)$, and more specifically b—y_2. Determine $\mathrm{tp}(x_1/y_1 by_2)$ and $\mathrm{tp}(x_2/y_1 ay_2)$ so that the amalgamation of (R, B_1) with (R, B_2) over (R, B_0) forces $\mathbb{A}$ into $\mathbb{H}$.

```
┌─────────────────────────┐
│  •x₂    ⊙y₁    ⊙y₂      │
│                 │       │
│  •x₁    •a     •b       │
└─────────────────────────┘
```

It remains to determine $\mathrm{tp}(x_1/a)$ and $\mathrm{tp}(x_2/b)$ so that the factors (R, B_1) and (R, B_2) embed in $\mathbb{H}$. As r is a ramsey type over B_1 we need only concern ourselves with (R, B_2). We first determine $\mathrm{tp}(x_1/a)$ by amalgamating $(x_1, \{by_2\})$ with $(\emptyset, B_2)$ over $(\emptyset, \{by_2\})$, and we then form (R, B_2) by amalgamating $(R, \{ay_2\})$ with (x_1, B_2) over $(x_1, \{ay_2\})$. The factor $(R, \{ay_2\})$ is available since $\{ay_2\} \simeq I_2$. □

LEMMA 4.23. *Let P be a 1-type over $K \simeq K(2)$ realized in $\mathbb{H}$, and let $\mathbb{H}^P = (^P K, K^\perp)$. Then $\mathbb{H}^P$ is ample.*

PROOF. In view of Proposition 10 this means that $\mathbb{H}$ embeds every $\mathbb{H}$-constrained 1-type over $K(2) + I_k$ for all k, which follows from Lemma 4.22 by induction on k. □

LEMMA 4.24. *Let (x, K_i) $(i = 1, 2)$ be 1-types over $K_i \simeq K(2)$ which are realized in $\mathbb{H}$. Then there is a 1-type (x, K) over $K \simeq K(2)$ such that the extensions $(x, K_1 + K)$, $(x, K_2 + K)$ both embed in $\mathbb{H}$.*

PROOF. We use the following amalgamation:

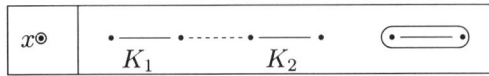

The factor omitting x is afforded by Propositions 10 and 11. The other factor comes from an amalgamation whose factors are obtained from Lemma 4.22. □

LEMMA 4.25. *Let $\mathbb{H}$ be an ample homogeneous 2-graph which embeds N 1-types over $K(2)$. Assume that Proposition 12 applies to every ample homogeneous 2-graph*

4.7. STEP 1, PROPOSITIONS 12 AND 13: REALIZATION OF 1-TYPES

which embeds at most N 1-types over $K(2)$. Then there is a ramsey type for $\mathbb{H}$ over $\infty \cdot K(2)$.

PROOF. By Lachlan's ramsey argument it suffices to show that every $\mathbb{H}$-constrained 1-type over $k \cdot K(2)$ (k arbitrary) embeds in $\mathbb{H}$. We prove this simultaneously for all ample homogeneous 2- graphs with at most N 1-types over $K(2)$, by induction on k. This is a direct application of Proposition 12 for this class of 2-graphs, using derived 2-graphs of the form $(^PK, K^\perp)$ with $K \subseteq H_2$, $K \simeq K(2)$. □

We now place ourselves in the following situation until the completion of the proof of Proposition 12:

> $\mathbb{H}$ is an ample homogeneous 2-graph realizing exactly N 1-types over $K(2)$, and Proposition 12 is assumed to hold for all ample homogeneous 2-graphs realizing fewer than N 1-types over $K(2)$.

We shall use the following notation. If P, Q are 1-types over $K(2)$ then $P + Q$ denotes the 1-type over $2 \cdot K(2)$ whose restrictions to the components of $2 \cdot K(2)$ are P and Q.

LEMMA 4.26. *Let P, Q be 1-types over $K(2)$ which are realized in $\mathbb{H}$. If $Q + Q$ is realized in $\mathbb{H}$, then $P + Q$ is realized in $\mathbb{H}$.*

PROOF. Let $\mathbb{H}^P = (^PK, K^\perp)$, $\mathbb{H}^Q = (^QK, K^\perp)$, where $K \subseteq H_2$, $K \simeq K(2)$. By Lemma 4.22 the cross types of $\mathbb{H}^P$ and $\mathbb{H}^Q$ coincide with those of $\mathbb{H}$. If $\mathbb{H}$ omits $P + Q$ then each of these 2-graphs realizes fewer than N 1-types over $K(2)$ and hence by our standing hypothesis, Proposition 12 applies to these 2-graphs.

Let r be a ramsey type for $\mathbb{H}^Q$ over $\infty \cdot K(2)$ and let R be an r-ramsey graph on two vertices x_1, x_2. We shall now describe a particular amalgamation diagram (pictured below) which forces an embedding of $P + Q$ into $\mathbb{H}$.

Let $A_0 = I + K$ with $I = \{a_1, a_2\} \simeq I_2$, and with $K \simeq K(2)$. For $i = 1, 2$ let $A_i = A_0 \cup \{b_i\}$ with $a_i \frown b_i$. We shall construct factors (R, A_1) and (R, A_2) in $\mathbb{H}$ agreeing on (R, A_0), in such a way that in their amalgam one of the following will necessarily be isomorphic to $P + Q$:

$$(\{x_1\}, I \cup \{b_1, b_2\}); \quad (\{x_2\}, \{b_1, b_2\} \cup K);$$

according as $b_1 \perp b_2$ or $b_1 \frown b_2$. In other words we require (in an obvious sense):

$$\mathrm{tp}(x_1/a_1 b_1) = \mathrm{tp}(x_2/b_1 b_2) = P;$$

$$\mathrm{tp}(x_1/a_2 b_2) = \mathrm{tp}(x_2/K) = Q.$$

We also fix a 1-type Q^* over $K(2)$ realized in both $\mathbb{H}^P$ and in $\mathbb{H}^Q$ (by Lemma 4.24) and we require $\mathrm{tp}(x_1/K) = Q^*$.

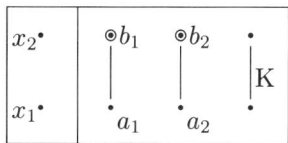

It remains to determine $\mathrm{tp}(x_2/a_1 a_2)$ so that both factors embed in $\mathbb{H}$. One obvious constraint is easily met:

$$\mathrm{tp}(x_2/a_2 b_2) \text{ is realized in } \mathbb{H}^Q.$$

If tp(x_2/a_2) is chosen so that this constraint is met then the factor (R, A_2) will embed in $\mathbb{H}^Q$ and *a fortiori* in $\mathbb{H}$ for any choice of tp(x_2/a_1), since the 1-types over $K(2)$ involved in (R, A_2) are all realized in $\mathbb{H}^Q$.

It remains to determine tp(x_2/a_1) so that (R, A_1) embeds in $\mathbb{H}$, and for this we simply amalgamate the subfactors $(R, \{b_1, a_2\} \cup K)$ and (x_1, A_1) over $(x_1, \{b_1, a_2\} \cup K)$. The first of these subfactors already embeds in $\mathbb{H}^Q$ since r is a ramsey type for H^Q, and the second is of the form $(x, I_1 + 2 \cdot K(2))$ with $(x, 2 \cdot K(2)) \simeq P + Q^*$, which may be seen to embed in $\mathbb{H}$ by applying Lemma 4.22 to $\mathbb{H}^P$. □

LEMMA 4.27. *Let P be a 1-type over $K(2)$ such that $P + P$ is not realized in $\mathbb{H}$. Then there is a cross type p such that for every k, for every 1-type Q_1 over I_k, and for every 1-type Q_2 over $K = \{a, b\}$ with a—b whose restriction to a is p, the 1-type $P + Q_1 + Q_2$ is realized in $\mathbb{H}$.*

PROOF. Suppose on the contrary that for each cross type p we have a counterexample k_p, $Q_1(p)$, $Q_2(p)$. Let $Q = \oplus_p Q_1(p)$, $k = \sum_p k_p$. Let m be the number of cross types in $\mathbb{H}$. Amalgamate $(x, I_m + I_k + K(2))$ with $(\emptyset, [\{y\}, I_m] + I_k + K(2))$ where tp(x/I_k) = Q, tp($x/K(2)$) = P, and tp(x/I_m) is made up of all the cross types. The second factor embeds in $\mathbb{H}$ since H_2 embeds all elements of $\mathcal{A}(0)$. Take $p = $ tp(xy) in the amalgam to arrive at a contradiction. □

LEMMA 4.28. *Let P be a 1-type over $K(2)$ such that $P + P$ is not realized in $\mathbb{H}$ and let r be a ramsey 2-type for $\mathbb{H}^P$ over I_∞, p a cross type. There is a 1-type Q over $K(2) = [a$—$b]$ whose restriction to a is p, such that for any finite configuration $\mathbb{A} = (R, I_k)$ with R r-ramsey, and any $x_1 \in R$, $\mathbb{H}$ embeds $(R, I_k + K)$ with:*

$$(R, I_k) \simeq \mathbb{A}, \quad (x_1, K) \simeq Q, \quad (x, K) \simeq P \text{ for all } x \in R - \{x_1\}.$$

PROOF. Suppose on the contrary that for each cross type q the 1-type Q over $K(2)$ represented by $[p$—$q]$ does not have the desired property, so that we have counterexamples $R(q), I_{k_q}, x_1(q)$. Take (R, I_k, x_1) isomorphically embedding each configuration $(R(q), I_{k_q}, x_1(q))$. To get a contradiction it suffices to amalgamate $(R, I_k + \{a\})$ with $(R - \{x_1\}, I_k + K)$, where $K = [a$—$b]$. Here $(R - \{x_1\}, I_k + K)$ embeds in $\mathbb{H}$ since $(R - \{x_1\}, I_k)$ embeds in $\mathbb{H}^P$. □

Now we shall prove Proposition 12.

PROOF. By Lemma 4.26, a counterexample $\mathbb{H}$ will fail to embed $P + P$ for some 1-type P over $K(2)$ which does embed in $\mathbb{H}$. Let N be the number of 1-types over $K(2)$ realized in $\mathbb{H}$, which we take to be minimized over all counterexamples $\mathbb{H}$. Let P be of the form p_1—p_2, let r be ramsey for $\mathbb{H}^P$ over I_∞, and let R be r-ramsey of order 2, $R = \{x_1, x_2\}$.

Let $A_0 = \{b_1, b_2\} + K \simeq I_2 + K(2)$, $A_i = A_0 \cup \{a_i\}$ for $i = 1, 2$ with a_i—b_i. It suffices to amalgamate (R, A_1) with (R, A_2), where:

$$\text{tp}(x_1/a_i b_i) = P \text{ for } i = 1, 2;$$

$$\text{tp}(x_2/K) = P; \quad \text{tp}(x_2/a_1) = p_1; \quad \text{tp}(x_2/a_2) = p_2.$$

4.7. STEP 1, PROPOSITIONS 12 AND 13: REALIZATION OF 1-TYPES

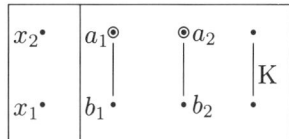

So our problem is to find compatible forms of these two factors in $\mathbb{H}$. We choose a cross type p according to Lemma 4.27 and then we choose a 1-type Q over $K(2)$ according to Lemma 4.28, and we shall require in addition:

$$\text{tp}(x_1/K) = Q.$$

Our construction will be based on the following claim:

(∗) There is a cross type s_1 such that for any k and any configuration $\mathbb{A}$ of the form (R, I_k), the configuration $(R, I_k + K_1 + K_2)$ embeds in $\mathbb{H}$, where $(R, I_k) \simeq \mathbb{A}$, $K_i = [a_i \mathrel{{-}} b_i] \simeq K(2)$, and:

$$\begin{array}{ll} \text{tp}(x_1/K_1) = Q; & \text{tp}(x_1/K_2) = P; \\ \text{tp}(x_2/K_1) = P; & \\ \text{tp}(x_2/a_2) = p_1; & \text{tp}(x_2/b_2) = s_1. \end{array}$$

We make a similar claim (∗∗) concerning the analogous situation with p_1 replaced by p_2; call the corresponding cross type s_2. The definition of (R, A_1) and (R, A_2) can be completed on the basis of these two claims by taking:

$$\text{tp}(x_2/b_1) = s_1; \quad \text{tp}(x_2/b_2) = s_2.$$

So it suffices to check (∗), as the treatment of (∗∗) is similar. We suppose accordingly that for all choices of cross types $s = s_1$ there are counterexamples k_s, $\mathbb{A}_s = (R, I_{k_s})$, to (∗). Embed all of these into a configuration (R, I_k). Then to reach a contradiction it will be sufficient to amalgamate $(R, I_k + K_1 + \{a\})$ with a configuration $(x_1, I_k + K_1 + K_2)$ where $K_2 = [a \mathrel{{-}} b]$ and:

$$\text{tp}(x_1/K_1 + K_2) = Q + P;$$

$$\text{tp}(x_2/K_1 + (a)) = P + p_1.$$

The factor $(R, I_k + K_1 + \{a\})$ embeds in $\mathbb{H}$ by the choice of Q, and the factor $(x_1, I_k + K_1 + K_2)$ embeds in $\mathbb{H}$ by the choice of p. □

Now we can prove Proposition 13 quickly. The assertion is that an ample homogeneous 2-graph $\mathbb{H}$ will embed all $\mathbb{H}$-constrained 1-types over P_3.

PROOF. By Lemma 4.25 there is a ramsey 2-type r over $\infty \cdot K(2)$ relative to $\mathbb{H}$. Let R be r-ramsey of order 2, $R = \{x_1, x_2\}$. Let $A_0 = \{a_1, a_2\} \simeq I_2$, $A_i = A_0 \cup \{b_i\}$, with $a_1, a_2 \mathrel{{-}} b_1$ and $a_2 \mathrel{{-}} b_2$. Choose $\text{tp}(x_1/b_1 a_2 b_2)$, and $\text{tp}(x_2/a_1 b_1 b_2)$ so that after amalgamating (R, A_1) and (R, A_2) over (R, A_0), one of these will be the desired 1-type over P_3. As $A_2 \simeq I_1 + K(2)$, it suffices to embed (R, A_1) in $\mathbb{H}$ in such a way that each $\text{tp}(x_i/a_2 b_2)$ is realized in $\mathbb{H}$. This is already the case for $i = 1$, but $\text{tp}(x_2/a_2)$ remains to be chosen.

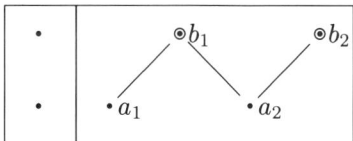

Form $A = A_1 \cup A_2$ with no new edges. Amalgamate:
$$(\emptyset, A) \text{ and } (x_2, \{a_1b_1b_2\}) \text{ over } (\emptyset, \{a_1b_1b_2\})$$
to form (x_2, A) In particular $\operatorname{tp}(x_2/a_2b_2)$ is realized in $\mathbb{H}$. Now we may amalgamate $(R, \{b_1a_2\})$ with (x_2, A_1) over $(x_2, \{b_1a_2\})$ to form (R, A_1) of the desired type. □

4.8. Step 2. Theorem 4.8: a, b, K

In the present section n is fixed and we assume Theorem 4.2 holds for $n-1$. We consider the configuration $\{a\} \cup K$ with $K \simeq K(n)$ and a linked to a unique point b of K. Our claim is:

aK belongs to any amalgamation class containing $\mathcal{A}(n)$

We use the following amalgamation. Let $A_o = (K_1 + K_2) \cup \{a_o\}$ where $K_1 \simeq K(n-1)$, $K_2 \simeq K(n-2)$, $a_o \perp K_1$, and a_o is linked to a unique element $b_2 \in K_2$. Let $A_1 = A_o \cup \{c_1\}$, $A_2 = A_o \cup \{c_2\}$, where c_1 is linked to K_2 and to a unique element b_1 of K_1, while c_2 is linked to all of $K_1 + K_2$, and $c_1, c_2 \perp a_o$. We shall also consider variants of A_o which differ only by the introduction of some additional edges connecting b_1 to points of K_2. For any variant A'_o of A_o we may consider the corresponding variants A'_1, A'_2 of A_1 and A_2. Notice that amalgamation of such A'_1 and A'_2 over A'_o will result in one of $c_1c_2K_1$ or $c_1c_2K_2a_o$ being isomorphic with aK.

Our objective is to find suitable factors A'_1, A'_2 in $\mathcal{A}$. We form A'_2 by amalgamating $K_1c_2a_o$ with $(K_1 - \{b_1\})K_2c_2a_o$ over $(K_1 - \{b_1\})c_2a_o$. The first factor is of the form $K(n) + I_1$ and the second factor omits $K(n)$, so both of these are found in $\mathcal{A}$. If b_1K_2 is complete then $b_1c_2K_2a_o \simeq aK$ and we are done. Assume therfore that b_1K_2 is not complete. Then the corresponding factor A'_1 omits $K(n)$ and hence belongs to $\mathcal{A}$. □

4.9. Step 6. Theorem 4.9.n: extending direct sums

The theorem asserts that $\mathcal{A}(n)$ entails any finite graph omitting $K(n+1)$ that is a one point extension of a disjoint sum of graphs in $\mathcal{A}(n)$.

In our proof we shall assume Propositions 11 and 13 as well as Theorem 4.3 and Theorem 4.8 for n.

PROOF. Fix H a homogeneous graph embedding all graphs in $\mathcal{A}(n)$, and a graph $A = \oplus A_i$ with $A_i \in \mathcal{A}(n)$, as well as an extension of A by one point x, omitting $K(n+1)$. Fix $a \in H$ and let $\mathbb{H} = (a', a^\perp)$. This is ample by repeated applications of Proposition 11, and we shall show that (x, A) embeds in $\mathbb{H}$. By Theorem 4.3 it suffices to show that (x, A_i) embeds in $\mathbb{H}$ for each i. For $A_i \simeq K(n)$ this follows from Theorem 4.8 and induction on n. For $A_i = P_3$ this reduces by Theorem 4.3 to the case $A_i = K(2)$. In H we are trying to embed certain configurations $\{a, x\} \cup A_i$. If $\{x\} \cup A_i \simeq K(3)$ then Theorem 4.8 applies, and if not then all cases fall under Proposition 11. □

4.10. Step 3. Theorem 4.6

Our hypothesis throughout this section and the next is that Propositions 11 and 13 are known, as well as Theorem 4.4.n_0 for $n_0 < n$ and Theorem 4.8 for n. We derive Theorem 4.6 for n, which says:

$$\mathcal{A}(n) \implies \mathcal{A}(n) + \mathcal{A}(n).$$

We also derive a useful result which we did not record explicitly in the outline of the proof, Lemma 4.30. We shall use this lemma also in the following section, so one should notice that this is legitimate.

LEMMA 4.29. $\mathcal{A}(n) \implies I_1 + \mathcal{A}(n)$

PROOF. That $\mathcal{A}(\emptyset) \implies I_1 + \mathcal{A}(\emptyset)$ is contained in Proposition 11. It remains to show that $\mathcal{A}(n) \implies I_1 + K(n)$.

Let $K_1 \simeq K(n-1)$, $K_2 \simeq K(n-2)$, and fix $b \in K_1$. Let $A_0 = K_1 \cup (K_2 + \{a\})$ where $K_1 - \{b\} \perp (K_2 + \{a\})$ and the type of b over $K_2 = \{a\}$ will be specified later. Let $A_i = A_0 \cup \{a_i\}$ for $i = 1, 2$ where a_1 is linked to $K_1 \cup K_2$ and a_2 is linked to K_2. In an amalgam of A_1, A_2 one of $a_2 a_1 K_1$ or $a a_1 a_2 K_2$ will be of the form $I_1 + K(n)$. Furthermore the factor A_2 omits $K(n)$, so is certainly a consequence of $\mathcal{A}(n)$. The factor A_1 may be formed by amalgamating $a_1 K_1$ with $A_1 - \{b\}$; the first subfactor is a copy of $K(n)$, and the second omits $K(n)$. □

LEMMA 4.30. *Suppose that $K \simeq K(n)$, A is a finite graph containing K, $b \in K$, and $A^* = A - \{b\}$ omits $K(n)$. Then $\mathcal{A}(n) \implies A$.*

PROOF. Let H be a homogeneous graph embedding all graphs in $\mathcal{A}(n)$. Let $a \in H$, and $\mathbb{H} = (a', a^\perp)$. Then $\mathbb{H}$ is ample. It suffices to show that (b, A^*) embeds in $\mathbb{H}$. By Theorem 4.4.$(n-1)$ it suffices to show that (b, K_0) embeds in $\mathbb{H}$ for K_0 a complete subgraph of A^*, or in other words that the graphs of the form $\{a, b\} \cup K_0$ with $a - b$ and $a \perp K_0$ embed in H. This follows from Theorem refABK.n, or induction. □

We prove Theorem 4.6: $\mathcal{A}(n) \implies \mathcal{A}(n) + \mathcal{A}(n)$.

PROOF. Let $A_1, A_2 \in \mathcal{A}(n)$. We have to show that $\mathcal{A}(n) \implies A_1 + A_2$. If $A_1 \simeq P_3$ then either Lemma 4.30 applies, or $A_1 + A_2$ is contained in $P_3 + P_3$, and Proposition 11 handles this case.

Accordingly we may take $A_i \simeq K(n)$, for $i = 1, 2$. Let $U; V_1, V_2; W$ be respectively copies of $K(n)$; $K(n-1)$ twice; $K(n-2)$. Let A_0 be derived from $U + V_1 + V_2 + W$ by inserting two edges, linking a vertex in U to a vertex in V_1 and a vertex in V_2. Let $A_i = A_0 \cup \{a_i\}$ where a_i is linked to $V_i + W$. Amalgamation of A_1 with A_2 produces a copy of $K(n) + K(n)$.

Since A_1, A_2 are isomorphic, it suffices to check that A_1 embeds in H, assuming that $K(n) + K(n)$ does not. Let b, c be the vertices of U and V_1 which are linked. Then it suffices to amalgamate $A_1 - \{c\}$ with $A_1 - \{b\}$ to get $K(n) + K(n)$ or A_1. Each of these last two factors contains a unique copy of K_n, to which Lemma 4.30 applies. □

4.11. Step 4. Theorem 4.7

Our hypothesis throughout this section and the next is that Propositions 11 and 13 are known, as well as Theorems 4.6 and 4.8 for n, and Theorem 4.4.n_0 for $n_0 < n$. In particular Lemma 4.30 of the previous section holds. We derive Theorem 4.7 for n, concerning the embedding of 1-types over graphs which are disjoint sums of two complete graphs. This is the final step in the proof of the classification theorem, and requires detailed amalgamation arguments which however have a somewhat less specialized flavor than most.

The proof of Theorem 4.7 for n occupies the remainder of this section. We consider a fixed counterexample to Theorem 4.7 for n, consisting of a homogeneous

2-graph $\mathbb{H}$ and a 1-type $\mathbb{A}$ over $K_1 + K_2$, with K_1, K_2 complete graphs of order at most n, such that the restriction of $\mathbb{A}$ to K_i embeds in $\mathbb{H}$ for $i = 1, 2$, but $\mathbb{A}$ does not embed in $\mathbb{H}$. As we are assuming that n is minimal, we may suppose that $K_1 \simeq K(n)$. We may also assume that our counterexample has also been chosen to minimize $|K_2|$. Finally, in the very important case in which $K_2 \simeq K(n)$ assume that the example has also been chosen to minimize the number of types realized in $\mathbb{H}$ over $K(n)$.

We consider three cases:

Case 1. $|K_2| < n$, and there is a complete graph K of order less than n such that some 1-type Q over K is not realized in $\mathbb{H}$.

Case 2. $|K_2| < n$, and for all complete graphs K of order less than n and all 1-types Q over K, Q is realized in $\mathbb{H}$.

Case 3. $K_1 \simeq K_2 \simeq K(n)$.

We have encountered the type of argument needed for Case 3 in the proof of Proposition 12, which dealt with the possibility $K_1 \simeq K_2 \simeq K(2)$. In the first two cases we can proceed in a fairly direct manner.

Case 1. $|K_2| < n$, and there is a 1-type Q over a complete graph K of order less than n which is not realized in $\mathbb{H}$.

DEFINITION 4. Fix Q, K as described with $|K|$ minimal. Fix $a_1 \in K_1$, $a_2 \in K_2$, $b_1, b_2 \in K$, and let K_1^*, K_2^*, K^* be $K_1 - \{a_1\}$, $K_2 - \{a_2\}$, and $K - \{b_1, b_2\}$ respectively. Let $A_0 = K_1^* + K_2^* + K^*$, $A_i = A_0 \cup \{c_i\}$ for $i = 1, 2$ with: c_i linked to K_i^* and K^* for $i = 1, 2$.

Applying Theorem 4.4.$(n-1)$, let r be a ramsey type for $\mathbb{H}$ over the $\mathcal{A}(n-1)$-generic graph. Let R be r-ramsey on two vertices x_1, x_2. We intend to amalgamate (R, A_1) with (R, A_2) over (R, A_0) in such a way that one of the following will be forced to occur:

$$(x_1, K^* \cup \{c_1, c_2\}) \simeq Q;$$

$$(x_2, (K_1^* + K_2^*) \cup \{c_1, c_2\}) \simeq \mathbb{A}.$$

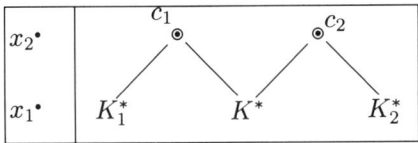

It remains to construct suitable factors (R, A_1) and (R, A_2) embedding in $\mathbb{H}$. The former is obtained by amalgamating:

$$(x_2, A_1) \text{ with } (R, (K^* \cup \{c_1\}) + K_2^*).$$

The subfactor (x_2, A_1) is produced by amalgamating $(x_2, \{c_1\} + K_1^* + K_2^*)$ with $(\emptyset, A_1)$ over $(\emptyset, \{c_1\} + K_1^* + K_2^*)$; $(\emptyset, A_1)$ is afforded by Lemma 4.30. The second subfactor is afforded by the choice of R, if it is $\mathbb{H}$-constrained. In fact we shall constrain $\text{tp}(x_i/K_2^*)$ for $i = 1, 2$ by requiring it to be the restriction of a type $\text{tp}(x_i/\{c_2\} \cup K_2^*)$ realized in $\mathbb{H}$. Since $|K^* \cup \{c_1\}| < |K|$, it is not necessary to impose additional constraints on $\text{tp}(x_2/K^*)$.

Now in view of the choice of R, to embed the corresponding factor (R, A_2) into $\mathbb{H}$ it suffices to check that it is $\mathbb{H}$-constrained, and this is immediate.

Case 2. $|K_2| < |K_1| = n$, and all 1-types over $K(n-1)$ are realized in $\mathbb{H}$.

DEFINITION 5. Let r and R be as in Case 1. Let $m = |K_2|$. Take copies U_1, U_2 of $K(n-2)$, and copies V_1, V_2 of $K(m-1)$. Let A_0 be $(U_1 + U_2 + V_1 + V_2) \cup \{b, c\}$ with b linked to U_1, c linked to V_1, and with no further edges with the possible exception of edges linking b and vertices in $U_2 \cup \{c\}$; these will be chosen later. Let $A_i = A_0 \cup \{a_i\}$ for $i = 1, 2$ with a_1 linked to U_1, U_2, and b, and with a_2 linked to U_2, V_2. We amalgamate (R, A_1) and (R, A_2) over (R, A_0), choosing the types involved so that one of the following necessarily is a copy of the desired 1-type $\mathbb{A}$:

$$(x_1, (U_1 \cup \{a_1, a_2\}) + (V_1 \cup \{c\})); \quad (x_2, (U_2 + V_2) \cup \{a_1, a_2, b\}).$$

It remains to find suitable factors (R, A_1) and (R, A_2) embedding in $\mathbb{H}$. Once (R, A_1) has been determined it will be easy to find a matching factor (R, A_2) in $\mathbb{H}$, since A_2 omits $K(n)$.

Thus it suffices to construct a suitable factor (R, A_1) in $\mathbb{H}$. We shall amalgamate:

$$(R, (\oplus_{i \leq 2} U_i \cup \{a_1\}) + (V_1 \cup \{c\}) + V_2)) \quad \text{with}$$
$$(R, V_2 + (U_1 \cup V_1 \cup \{a_1, b, c\})).$$

The former embeds in $\mathbb{H}$ by the choice of R. For the latter we amalgamate $(R, V_1 + (V_2 \cup \{a_1, b, c\}))$, which is afforded by the choice of R, with $(x_2, V_2 + (U_1 \cup V_1 \cup \{a_1, b, c\}))$, in which the type of bc still remains to be specified. So the final subfactor $(x_2, V_2 + (U_1 \cup V_1 \cup \{a_1, b, c\}))$, may be constructed by amalgamating $(x_2, V_1 + V_2 + (U_1 \cup \{a_1, b\}))$ with $(x_2, (V_1 \cup \{c\}) + V_2 + (U_1 \cup \{a_1\}))$. We shall check that these two configurations occur in $\mathbb{H}$.

$(x_2, (V_1 \cup \{c\}) + V_2 + (U_1 \cup \{a_1\}))$ is afforded by induction on n. To see that $(x_2, V_1 + V_2 + (U_1 \cup \{a_1, b\}))$ also occurs, let $p = \text{tp}(x_2/V_1)$, $\mathbb{H}^p = (^p V_2, V_2^\perp)$, and work in $\mathbb{H}^p$. By the minimality of the counterexample $\mathbb{H}, \mathbb{A}$, we have both $(x_2, U_1 \cup \{a_1, b\})$ and (x_2, V_1) in $\mathbb{H}^p$, and hence (again by minimality) $(x_2, (V_1 \cup \{c\}) + (U_1 \cup \{a_1\}))$ also embeds in $\mathbb{H}^p$, as required.

Case 3. $K_1 \simeq K_2 \simeq K(n)$.

LEMMA 4.31. *Let P_1, P_2 be two 1-types over $K(n)$ realized in $\mathbb{H}$. Then there is a 1-type Q over $K(n)$ which is realized in $\mathbb{H}^{P_1}$ and in $\mathbb{H}^{P_2}$.*

PROOF. Let K_1, K_2, K_3 be three disjoint copies of $K(n)$. Let $A_0 = K_1 \cup K_2$ with at most one extra edge joining some vertex $a \in K_1$ with some $b \in K_2$; in other words, $\text{tp}(ab)$ will be specified later. Let $\mathbb{A}_1 = (x, A_0)$ with $\text{tp}(x/K_i) = P_i$ for $i = 1, 2$, and let $\mathbb{A}_2 = (\emptyset, A_0 + K_3)$. An amalgamation of $\mathbb{A}_1$ with $\mathbb{A}_2$ will produce a suitable 1-type (x, K_3). It remains to be seen that $\mathbb{A}_1$ and $\mathbb{A}_2$ embed into $\mathbb{H}$.

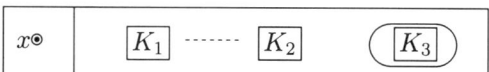

$\mathbb{A}_1$ can be obtained in some form by amalgamating $(x, A_0 - \{a\})$ with $(x, A_0 - \{b\})$. It remains to check that H_2 embeds the corresponding configuration $A_0 + K_3$. By Theorem 4.6 it suffices to check that H_2 embeds $K_1 \cup K_2$ with just one edge linking K_1 to K_2 (as usual, we really replace H_2 by $K_3^\perp$ momentarily). Let $c \in H_2$ and set $\mathbb{H}^* = (c', c^\perp)$. We shall show that $\mathbb{H}^*$ embeds $(a, A_0 - \{a\})$. As $A_0 - \{a\} = (K_1 - \{a\}) + K_2$, by the cases of Theorem 4.7.n already treated it suffices to prove that $\mathbb{H}^*$ embeds $(a, K_1 - \{a\})$ and (a, K_2). Reformulated in terms of H_2 these embeddings will follow from Lemma 4.30. □

LEMMA 4.32. *Let P, Q be 1-types over $K(n)$ realized in $\mathbb{H}$. If $P + P$ embeds in $\mathbb{H}$, then $P + Q$ embeds in $\mathbb{H}$.*

PROOF. Let Q^* be a 1-type over $K(n)$ which is realized in both $\mathbb{H}^P$ and $\mathbb{H}^Q$. Assume that $\mathbb{H}$ omits $P + Q$. As $\mathbb{H}^P$ realizes fewer types over $K(n)$ than $\mathbb{H}$ does, by induction Theorem 4.7 applies to $\mathbb{H}^P$. Hence Theorem 4.3 for n also applies to $\mathbb{H}^P$ (see the precise formulation of the inductive proof of Theorem 4.3 given earlier). Therefore there is a ramsey type r for $\mathbb{H}^P$ over the elements of $\mathcal{A}(n)$. Let $R = \{x_1, x_2\}$ be r-ramsey of order 2. Any $\mathbb{H}^P$-constrained 2-graph of the form (R, A) with $\mathcal{A}(n) \implies A$ will embed in $\mathbb{H}^P$, and a fortiori in $\mathbb{H}$.

Let $U; V_1, V_2; W$ be disjoint copies of $K(n); K(n-1); K(n-2)$ respectively. Let $A_0 = U + V_1 + V_2 + W$, $A_i = A_0 \cup \{a_i\}$ for $i = 1, 2$ with a_i linked to V_i and W. Observe that $\mathcal{A}(n) \implies A_1$ by Theorem 4.6 4.30.

We shall amalgamate configurations of the form (R, A_1) and (R, A_2) in $\mathbb{H}$ to force a realization of $P + Q$. We require:

(1): $\text{tp}(x_1/U; V_1 \cup \{a_1\}; V_2 \cup \{c_2\}) = Q^*; P; Q$ respectively;
(2): $\text{tp}(x_2/U; W \cup \{a_1, a_2\}) = P; Q$
(3): $\text{tp}(x_2/V_1 \cup \{a_1\})$ is realized in $\mathbb{H}^P$.

Suppose that we have constructed (R, A_2) embedding in $\mathbb{H}$, and satisfying:

(4): $(R, \{a_1\} \cup W)$ is $\mathbb{H}$-constrained.

We claim that (R, A_1) is then $\mathbb{H}^P$-constrained. $(R, U + (V_1 \cup \{a_1\}))$ is $\mathbb{H}^P$-constrained by (1–3). $(R, \{a_1\} \cup (W + V_2))$ is $\mathbb{H}$-constrained by (4), and hence also $\mathbb{H}^P$-constrained by the previous cases of Theorem 4.7. So to complete the argument it suffices to build (R, A_2) in $\mathbb{H}$ subject to (1–4).

Let $\text{tp}(x_2/V_1 \cup \{a_1\})$ be specified arbitrarily subject to (3), and the constraint on $\text{tp}(x_2/a_1)$ contained in (2). We shall refer to this more specific constraint as "constraint (3)" as well. We now construct a configuration (R, A_2) in $\mathbb{H}$ compatible with clauses (1–4) by amalgamating suitable factors $(R, A_2 - V_2)$ and (x_1, A_2). Here $(R, A_2 - V_2)$ is chosen so that $(R, \{a_1, a_2\} \cup W)$ is $\mathbb{H}$-constrained (taking $a_1 \perp a_2$ momentarily). This is easy to do.

We still have to describe (x_1, A_2). Let $P^* = Q^* + P \upharpoonright V_1$, as a type over $U + V_1$. The constraints on (x_1, A_2) are:

$$\text{tp}(x_1/U + V_1) = P^*;$$

$\text{tp}(x_1/\{a_1, a_2\} \cup W)$ is $\mathbb{H}$-constrained if $a_1 \perp a_2$;

$$\text{tp}(x_1/a_1) = P \upharpoonright a_1; \text{tp}(x_1/a_2 V_2) = Q.$$

Working in $\mathbb{H}^{P^*}$, which is also ample homogeneous embedding P and Q, we set $A = \{a_1, a_2\} \cup (W + V_2)$ and form (x_1, A) by amalgamating $(x_1, \{a_1\} + (\{a_2\} \cup V_2))$ with $(\emptyset, A)$. Both embed in $\mathbb{H}^{P^*}$, the latter by Lemma 4.30. □

NOTATION 3. *For the remainder of the proof we fix a 1-type P over $K(n)$ such that P embeds in $\mathbb{H}$ and $P + P$ does not embed in $\mathbb{H}$.*

LEMMA 4.33. *Let $K \simeq K(n)$, and let $A = K \cup J$ be a finite graph such that for some $a \in K$, $J \perp K - \{a\}$. If $J \cup \{a\}$ omits $K(n)$ then every $\mathbb{H}$-constrained 1-type Q over A is realized in $\mathbb{H}$.*

4.11. STEP 4. THEOREM 4.7

PROOF. We proceed by induction on $|J|$, starting at $J = \emptyset$. We may suppose that $n > 2$. Let $b, c \in K - \{a\}$, $d \in J$. Let J_1, J_2 be disjoint copies of $J - \{d\}$ and J, and let K_1, K_2 be disjoint copies of $K - \{b\}$ and $K - \{b,c\}$, arranged so that $J_1 \cup K_1$ and $J_2 \cup K_2$ are copies of the appropriate subgraphs of $J \cup K$.

Let $A_0 = (K_1 \cup J_1) \cup (K_2 \cup J_2)$ with $K_i \perp J_j$ for $i \neq j$. Let $c_1 \in K_1$ correspond to $c \in K$ and take $K_1 - \{c_1\} \perp K_2$. We shall determine $\mathrm{tp}(c_1/K_2)$ later. Let $A_i = A_0 \cup \{e_i\}$ for $i = 1, 2$ with:

$$(e_1, J_1 \cup K_1) \simeq (b, A - \{b,d\}); \quad (e_2, J_1 \cup K_1) \simeq (d, A - \{b,d\});$$
$$(e_1, J_2 \cup K_2) \simeq (b, A - \{b,c\}); \quad (e_2, J_2 \cup K_2) \simeq (c, A - \{b,c\});$$

and with e_1, e_2 linked to K_2 and unlinked to J_2. Let (x, A_1), (x, A_2) be taken so that after amalgamation one of the configurations $(x, \{e_1, e_2\} \cup K_i \cup J_i)$ is a realization of Q for $i = 1$ or 2. It remains to construct suitable factors (x, A_1), (x, A_2) in $\mathbb{H}$.

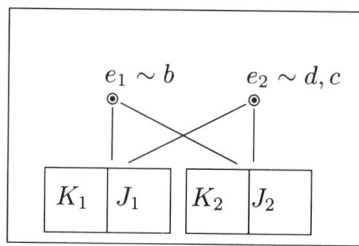

To construct (x, A_1), amalgamate $(x, A_1 - \{c_1\})$ with $(x, A_1 - K_2)$. Both are in any case $\mathbb{H}$-constrained, and we claim that both embed in $\mathbb{H}$. As $A_1 - \{c_1\}$ omits $K(n)$, $(x, A_1 - \{c_1\})$ embeds in $\mathbb{H}$. Now $A_1 - K_2 = \{e_1\} \cup (K_1 \cup J_1) + J_2$. Let $Q_0 = \mathrm{tp}(x/J_2)$. Working in $\mathbb{H}^{Q_0}$, it suffices to embed $(x, \{e_1\} \cup K_1 \cup J_1)$. Since $\{e_1\} \cup K_1 \cup J_1$ omits $K(n)$, this can be done by induction; it suffices to check that $(x, \{e_1\} \cup K_1 \cup J_1)$ is $\mathbb{H}^{Q_0}$-constrained. This is clear.

Now we claim that the corresponding factor (x, A_2) is $\mathbb{H}$-constrained, and that A_2 omits $K(n)$, so that (x, A_2) embeds in $\mathbb{H}$ by induction. In fact every clique with at least three vertices contained in A_2 is contained in one of the following:

$$J_1 \cup \{a_1\}; \ J_2 \cup \{a_1\}; \ K_2 \cup \{c_1\}; \ K_1 \cup \{e_2\}; \ K_2 \cup \{e_2\} \qquad \square$$

LEMMA 4.34. *Let $a \in K \simeq K(n)$. There is a cross type q such that for any 1-type Q over $K(n)$ whose restriction to a is q, and for any 1-type P^* over $K \cup J$ with $P^* \upharpoonright K = P$, if $J \perp K - \{a\}$, $\{a\} \cup J$ omits $K(n)$, and Q and P^* are realized in $\mathbb{H}$, then $Q + P^*$ is realized in $\mathbb{H}$.*

PROOF. Suppose not. For each cross type q we may fix a counter example $Q(q), J(q), P^*(q)$. As q varies let $K(q)$ be disjoint copies of $K - \{a\}$, and let $K^* = (\oplus_q K(q)) \cup \{a\}$ with $(a, K(q)) \simeq (a, K - \{a\})$. Let $A^* = (\oplus_q J(q)) \cup K$ with $(K, J(q))$ as in $P^*(q)$. Let $A_0 = (K^* - \{a\}) + A^*$, $\mathbb{A}_1 = (\emptyset, K^* \cup A^*)$, $\mathbb{A}_2 = (x, A_0)$ with:

$$\mathrm{tp}(x/K(q)) = Q \upharpoonright K(q); \quad \mathrm{tp}(x/K \cup J(q)) = P^*(q).$$

Amalgamation of $\mathbb{A}_1$ with $\mathbb{A}_2$ produces a contradiction.

To see that $\mathbb{A}_1$ embeds in H_2 it suffices to apply Lemma 4.30.

To see that $\mathbb{A}_2$ embeds in $\mathbb{H}$ let $Q^* = \oplus_q Q(q) \restriction K(q)$, and let $\mathbb{H}^* = \mathbb{H}^{Q^*}$. It suffices to show that (x, A^*) embeds in $\mathbb{H}^*$, which follows from the previous lemma applied to $\mathbb{H}^*$, as $(x, A_0 - K)$ is $\mathbb{H}^P$-constrained. □

Now let r be a Ramsey 2-type for $\mathbb{H}^P$ over $\mathcal{A}(n-1)$. Again fix $a \in K \simeq K(n)$.

LEMMA 4.35. *For any cross type q there is a 1-type Q over $K(n)$ whose restriction to a is q, such that for any finite configuration $\mathbb{A} = (R, A)$ with R r-ramsey, A omitting $K(n)$, and for any $x_0 \in R$, $\mathbb{H}$ embeds a configuration of the form $(R, A + K(n))$ with $(R, A) \simeq \mathbb{A}$, $\operatorname{tp}(x_0/K) = Q$, and $\operatorname{tp}(x/K) = P$ for $x \in R - \{x_0\}$.*

PROOF. Otherwise, for each type Q over $K(n)$ whose restriction to a is q, we have a counterexample $(R(q), A(q), x_0(q))$. Take (R, A, x_0) isomorphically embedding each of these, with $A = \oplus_q A(q)$. To get a contradiction it suffices to amalgamate $(R, A + \{a\})$ with $(R - \{x_0\}, A + K(n))$. The latter embeds in $\mathbb{A}$ since $(R - \{x_0\}, A)$ embeds in $\mathbb{A}^P$. □

Now we conclude the proof of Theorem 4.7 by forcing $P + P$ into $\mathbb{H}$.

DEFINITION 6.
P is the given 1-type over $K \simeq K(n)$.
$a, b \in K$.
r is ramsey for $\mathbb{H}^P$ over $\mathcal{A}(n-1)$.
$R = \{x_1, x_2\}$ is r-ramsey.
$U; V_1, V_2; W$ are copies of $K(n); K - \{a\}$ twice; $K - \{a, b\}$.
$A_0 = U + V_1 + V_2 + W$;
$A_i = A_0 \cup \{a_i\}$ with a_i linked to V_i and W.
q is a cross type afforded by Lemma 4.34;
Q is a 1-type over $K(n)$ afforded by Lemma 4.35 applied to q.

Now to force $P + P$ into $\mathbb{H}$ it suffices to amalgamate (R, A_1) with (R, A_2) where:

$$\operatorname{tp}(x_1/U) = Q; \quad \operatorname{tp}(x_1/V_i \cup \{a_i\}) = P \text{ for } i = 1, 2;$$

$$\operatorname{tp}(x_2/U) = P; \operatorname{tp}(x_2/W \cup \{a_1, a_2\}) = P \text{ apart from } \operatorname{tp}(a_1 a_2).$$

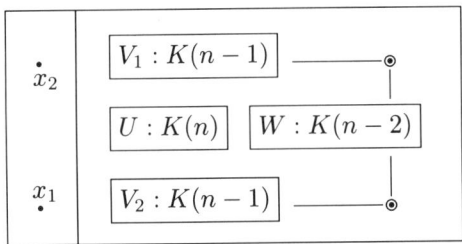

It must be checked that suitable factors (R, A_1) and (R, A_2) are available in $\mathbb{H}$. We claim that for any cross type p:

(*) there is a type P^* over $V_1 \cup \{a_1\}$, realized in $\mathbb{H}$, with $P^* \restriction a_1 = p$, such that for any $\mathbb{H}$-constrained configuration $\mathbb{A} = (R, U + (V_1 \cup \{a_1\} \cup J))$ with: $\operatorname{tp}(x_1/U) = Q; \operatorname{tp}(x_1/V_1 \cup \{a_1\}) = P; \operatorname{tp}(x_2/U) = P; \operatorname{tp}(x_2/V_1 \cup \{a_1\}) = P^*; J \perp (U + V_1); a_1 \perp U; \{a_1\} \cup J$ omitting $K(n)$ we have: $\mathbb{A}$ embeds in $\mathbb{H}$.

We shall apply (*) with $p = \operatorname{tp}(x_2/a_i)$, $i = 1, 2$, getting types P_1^*, P_2^*; it will then suffice to take $\operatorname{tp}(x_2/V_i \cup \{a_i\}) = P_i^*$ in $(R, A_1), (R, A_2)$ to arrive at compatible factors embedded in $\mathbb{H}$.

To prove $(*)$ suppose on the contrary that for every choice of P^* we have a counterexample $\mathbb{A}(P^*) = (R, U + (V_1 \cup \{a_1\} \bigcup J))$, where $R = R(P^*)$ and $J = J(P_1, P_2)$. Let $(R^*, V_1 \cup \{a_1\} \cup J)$ embed all $\mathbb{A}(P^*)$ with $J = \oplus_{P^*} J(P^*)$. Then we arrive at a contradiction by amalgamating:

$$(R^*, U + (\{a_1\} \cup J)) \text{ with } (x_1, U + (V_1 \cup \{a_1\} \cup J)).$$

The first factor embeds in $\mathbb{H}$ by the choice of Q, and the second factor embeds in $\mathbb{H}$ by the choice of p.

CHAPTER 5

Homogeneous directed graphs omitting I_∞

5.1. A catalog of homogeneous directed graphs

Our main goal in this Memoir is the complete classification of the homogeneous directed graphs. This chapter and the next are devoted to the study of homogeneous directed graphs in which any infinite induced directed subgraph contains an edge. In the present chapter we reduce the classification of these homogeneous directed graphs to five comparatively concrete propositions and one involved special case. The propositions will be proved in the next chapter, and the special case required will also be treated there.

We begin with a catalog. The notation will be explained afterward.

CATALOG 2. The homogeneous directed graphs.

I. Omitting some 2-type.
    1. I_n
    2. C_3, $\mathbb{Q}$, S(2), T^∞.

II. Imprimitive
    3. Wreathed (composite)
    4. $\hat{T}$, $T = I_1, C_3, \mathbb{Q}, T^\infty$
    5. $n * I_\infty$
    6. semigeneric

III. Exceptional
    7. S(3)
    8. $\mathcal{P}$
    9. $\mathcal{P}(3)$

IV. Free
    10. Γ_n
    11. $\mathcal{T}$-generic

In the first class we place those examples which are directed graphs only by courtesy, since they omit at least one of the nontrivial 2-types present in the language of directed graphs. I_n denotes the edgeless directed graph on n vertices. I_1 does double duty as a homogeneous tournament; the four other homogeneous tournaments are listed in the second class. For this we refer to Lachlan's original classification [57] or the exposition in [18].

5.1. A CATALOG OF HOMOGENEOUS DIRECTED GRAPHS

The next four classes exhaust the imprimitive examples, that is the ones with a nontrivial equivalence relation definable without parameters. This includes the rest of the finite examples, found in groups 3,4. For this classification we refer to [56] for the finite case, and [17] for the infinite case. Class 3 consists of the wreath product graphs obtained by composing a tournament and an edgeless graph in either of the two possible orders. As a matter of notation, the wreath product $I_n[T]$ is also denoted nT or $n \cdot T$, while $T[I_n]$ has no nickname. Class 4 consists of twisted versions of $T[I_2]$ which can be constructed systematically from a tournament T, and which happen to yield homogeneous directed graphs in four of the five possible cases in which T is itself a homogeneous tournament. The fifth example may be obtained from a complete n-partite graph (where possibly $n = \infty$) by generification: symmetric edges are oriented randomly. The sixth example is a restricted version of $\infty * I_\infty$ in which a parity restriction is placed on the edges between two pairs of two vertices, where the vertices in each pair are equivalent (unlinked). This example was found as a result of the classification project.

The last two graphs are afforded by free amalgamation classes according to Fraïssé's construction. There are two relevant notions of free amalgamation. One chooses a symmetric 2-type or a pair of antisymmetric 2-types, to be considered "neutral". An amalgamation diagram is then considered to be "freely amalgamated" if no identifications of vertices are made, and only neutral 2-types are used to complete the amalgam. With respect to this notion, the amalgamation indecomposable objects are those which omit the neutral type(s), that is: I_n in one case, tournaments in the other. A freely generated amalgamation class is one that consists of all the directed graphs which do not contain (as induced directed subgraphs, of course) any of a specified class of amalgamation indecomposable graphs. More concretely, Γ_n denotes the generic directed graph in which I_{n+1} does not embed, and if $\mathcal{T}$ is a class of finite tournaments, then $\Gamma(\mathcal{T})$ denotes the generic graph for which every induced subtournament embeds in one of the specified tournaments.

The three graphs listed as numbers 7-9 are indeed exceptional and will require a certain amount of extra attention in the course of our analysis. S(n) denotes the set of points lying at a rational angle θ on the unit circle, with n binary relations R_i ($i \in \mathbb{Z}/n\mathbb{Z}$) defined by:

$$R_i(a,b) \text{ iff the angle from } a \text{ to } b \text{ is in the range } (2\pi i/n, 2\pi(i+1)/n)$$

If we fix a point in S(n) and look at the structure on the remaining points, we get a shuffled n-tournament with two cross types between each pair of components. Hence it is easy to see that these structures are homogeneous, with n nontrivial 2-types.

The directed graph $\mathcal{P}$ is the generic (universal homogeneous) partial ordering. The homogeneous partial orderings were classified by Schmerl in [82], and we shall use that result later.

The last graph, $\mathcal{P}(3)$, was the last homogeneous directed graph discovered (1988). It is the only one omitted from the list of known examples which was given in [17]. Just as S(2) is derived from a shuffled 2-tournament with components of type $\mathbb{Q}$, $\mathcal{P}(3)$ is derived from a shuffled 3-directed graph with components of type $\mathcal{P}$. It will be described in detail in the next section, since it has not been described previously, and its homogeneity will be checked.

Our claim is that this list is complete. The proof occupies the rest of the present Memoir.

In addition to the notation introduced above, we use the notation $L(n)$ for a linear ordering (as a tournament) of order n, and P_n for an oriented path of order n. (Order refers to the number of vertices, not the "length".) We use much the same notation for types as in the case of tournaments, but we also need a symbol for the relation "x,y are unlinked", which we write: "$x \perp y$". For the most part we treat $\perp$ as an irreflexive relation. The only exception arises when we consider the possibility that "$\perp$" is an equivalence relation, where we naturally refer to its reflexive closure. Correspondingly $a^\perp$ will be the set of nonneighbors of a in a given directed graph.

We also use the notation $X+Y$ for the disjoint union of directed graphs X,Y, with no new edges, and $[X,Y]$ for the ordered disjoint union in which all pairs in $X \times Y$ are added as new edges.

If a directed graph A cannot be embedded in a directed graph H (as an induced directed subgraph – this is our standard convention) then we say that H omits A. When A can be embedded in H we say that H embeds A, and occasionally that H realizes A. Although we make no distinction between the two terms, for the sake of uniformity we lean toward the former, though there are certain occasions where A is an extension of a fixed configuration by a single additional point where we fall into the latter manner of speaking.

We shall speak of n-directed graphs, especially for $n=2$. By convention we shall allow at most three cross types between distinct components. This is harmless enough in the context of n-directed graphs derived from directed graphs, but undeniably awkward. We make this convention because we seem to need it in some of our proofs, though we believe the lemmas involved should hold more generally.

5.2. The graph $\mathcal{P}(3)$

Call a subset T of the generic partial order $\mathcal{P}$ dense if for any pair a,b of elements of $\mathcal{P}$ with $a \longrightarrow b$ we have an element $c \in T$ with $a \longrightarrow c \longrightarrow b$. Equivalently, any type realized in $\mathcal{P}$ over a finite subset A by an element not in A will be realized by an element of T. To see this, let a,b be two realizations of the given type with $a \longrightarrow b$ and then take c between a and b. In particular such a set is isomorphic with $\mathcal{P}$.

Construction.

Partition $\mathcal{P}$ into three dense subsets P_i ($i \in \mathbb{Z}/3\mathbb{Z}$). Let $\mathbb{P} = (P_0, P_1, P_2)$ be the 3-directed graph induced by the structure on $\mathcal{P}$. Identify the cross types $\longrightarrow, \perp, \longleftarrow$ with $-1, 0, 1$ respectively in $\mathbb{Z}/3\mathbb{Z}$. In particular $\operatorname{Sym}(2)$ acts on the cross types by multiplication by -1. Let $\mathbb{H}$ be the variant of $\mathbb{P}$ obtained by shifting the cross types in $P_i \times P_j$ by $j-i$. (Since $-k+(i-j) = -(k+(j-i))$ this definition respects the action of $\operatorname{Sym}(2)$.) Since $\mathbb{P}$ is clearly homogeneous, its variant $\mathbb{H}$ is also homogeneous.

Let P be the underlying set of $\mathcal{P}$. We let $\mathcal{P}(3)$ be the directed graph on the set $P \cup \{a\}$ characterized by $(a^\perp, a', 'a) = \mathbb{H}$. Our claim is that $\mathcal{P}(3)$ is homogeneous. Since it becomes homogeneous after fixing the point a, it suffices to check that the automorphism group is transitive, or equivalently that for all $b \in P$, the 3-directed graph $(b^\perp, b', 'b)$ (computed in $\mathcal{P}(3)$) is isomorphic with $\mathbb{H}$.

We introduce the following notation. Let S be a partial order, and $a \in S$. Let $S_i = a^i$ for $i \in \mathbb{Z}/3\mathbb{Z}$, identifying group elements and 2-types. Let S^a be the same set S equipped with a relation $<^a$ derived from $<$ by shifting the types on $S_i \times S_j$ by $i - j$ when they do not involve the element a, and by taking a^i in the sense of $<^a$ to be a^{-i} in the sense of $<$.

LEMMA 5.1. *Let S be a partial order and $a \in S$. Then:*
(1) *S^a is a partial order;*
(2) *$S^{aa} = S$;*
(3) *If $S \simeq \mathcal{P}$ then $S^a \simeq \mathcal{P}$.*

PROOF. (2) is immediate and (1) is easily checked. For (3), if $A \leq S^a$ is finite and $B = A \cup \{b\}$ is a partial order extending A, it suffices to check that B embeds in S^a over A. Now B^a certainly embeds in S over A^a, so $B = B^{aa}$ embeds in S^a over $A^{aa} = A$. □

Now for $b \in \mathcal{P}$ let $\mathbb{H}^b$ be the 3-directed graph $(b^\perp, b', 'b)$ computed inside $\mathcal{P}(3)$. Let $\mathbb{P}^b$ be derived from $\mathbb{H}^b$ by shifting the types realized in $b^i \times b^j$ by $i - j$. We claim that $\mathbb{H}^b \simeq \mathbb{H}$, or, equivalently:

LEMMA 5.2. $\mathbb{P}^b \simeq \mathbb{P}$.

PROOF. Let b be in P_i, and let c_l ($l = 1, 2$) lie in $b^{j_l} \cap P_{k_l}$, from the point of view of $\mathbb{P}$. Assume that b, c_1, c_2 are distinct. Let $p = \text{tp}(c_1, c_2)$, also computed in $\mathbb{P}$. Then in $\mathbb{H}$ the situation is as follows:

$$\text{tp}(c_1, c_2) = p + k_2 - k_1; \text{tp}(bc_l) = j_l + k_l - i;$$

so in $\mathbb{P}^b$ we find $\text{tp}(c_1, c_2) = p + k_2 - k_1 + (j_1 + k_1 - i) - (j_2 + k_2 - i) = p + j_1 - j_2$. Similarly $\text{tp}(ac_l)$ computed in $\mathcal{P}^b$ comes out to $-j_l = -\text{tp}(bc_l)$ as computed in $\mathbb{P}$. Thus the directed graph underlying $\mathbb{P}^b$ is exactly $\mathcal{P}^b$, and $\mathcal{P}^b \simeq \mathcal{P}$ by the lemma.

Now we need only check that b^j is dense in $\mathcal{P}^b$ for each j, when b^j is computed in $\mathcal{P}$ (for this purpose we may simply ignore a). More specifically we claim that for any j, k_1, k_2 the set $b^j \cap P_{k_1}$ is dense in the set $b^j \cap P_{k_2}$, where the sets are computed in $\mathbb{P}$ but at the same time the notion of density is taken in $\mathcal{P}^b$. Now our computation above shows that up to a permutation of the components the structure $(b^j \cap P_0, b^j \cap P_1, b^j \cap P_2)$ is the same whether viewed in $\mathbb{P}$ or in $\mathbb{P}^b$. So our claim is evident. □

We conclude that $\mathcal{P}(3)$ is homogeneous. Notice that for $a \in \mathcal{P}(3)$, $a^\perp, a', 'a$ are all isomorphic to $\mathcal{P}$. Of course the same is true in $\mathcal{P}$, so one should also notice that $\mathcal{P}(3)$ embeds an oriented 3-cycle, for example any three incomparable points of $\mathcal{P}$ lying in three distinct components of our initial partition will become a 3-cycle in $\mathcal{P}(3)$.

5.3. The theorem

THEOREM 5.3. *Let H be a primitive homogeneous directed graph omitting I_∞. Then H is isomorphic to one of the following: I_n, C_3, $\mathbb{Q}$, $S(2)$, $S(3)$, Γ_n with n finite.*

We shall use an inductive argument with some complications. We replace the theorem by a theorem in two parts, to be proved by simultaneous induction, where one part is a reformulation of the desired result, and the other part is inserted to close the circle and keep the induction moving.

We first introduce the important class $\mathcal{A}_n$ of "generators for Γ_n".

DEFINITION 5.4.   1. $\mathcal{A}_1 = \{[I_1, C_3]\}$;
2. $\mathcal{A}_n = \{(I_1 + L(2)), [I_1, I_2], I_n\}$ for $n > 1$.

The main point in this definition is that for $n > 1$ any homogeneous directed graph embedding the first two directed graphs listed in $\mathcal{A}_n$ will necessarily be primitive.

THEOREM 5.5 (Main Theorem (Part 1)). *For $n \geq 1$:*

(MT1.n) *If an amalgamation class $\mathcal{A}$ contains $\mathcal{A}_n$ and omits I_{n+1}, then $\mathcal{A}$ consists of exactly the directed graphs omitting I_{n+1}.*

The second part of the Main Theorem is concerned with homogeneous 2-directed graphs. By an abuse of terminology, a 2-directed graph $\mathbb{A}$ will be called a *1-type over the directed graph A* if $A_2 \simeq A$ and $|A_1| = 1$. We emphasize the convention established earlier, that in a 2-directed graph there are at most three cross types.

THEOREM 5.6 (Main Theorem (Part 2)). *For $n \geq 1$:*

(MT2.n) *Let $\mathbb{H}$ be a homogeneous 2-directed graph whose first component H_1 omits I_∞, and whose second component H_2 is isomorphic with Γ_n. Assume that $\mathbb{H}$ embeds all 1-types over $L(2)$ and over I_n, and that there are at least two cross types. Then $\mathbb{H}$ embeds all finite 2-directed graphs $\mathbb{A}$ for which A_1 is linear and A_2 omits I_{n+1}.*

In quoting this result, we shall usually not include the hypothesis on the number of cross types explicitly. It is needed for Lachlan's Ramsey argument, and without this hypothesis we could let H_1 be any finite homogeneous directed graph, such as I_1.

The form of the proof of the Main Theorem is *in spirit* as follows:

(1) MT 1.(n-1) & MT 2.(n-1) $\implies$ MT 1.n
(2) MT 1.n & MT 2.(n-1) $\implies$ MT 2.n

However the technical Proposition 18 below also depends on a parameter n, and is itself proved by induction. Let us refer to this statement for the moment as 18_n. The exact relations established are the following:

(1') MT 1.(n-1) & MT 2.(n-1) & 18_n $\implies$ MT 1.n;
(2) MT 1.n & MT 2.(n-1) $\implies$ MT 2.n;
(3) 18_{n-1} & MT 1.(n-2) & MT 2.(n-2) $\implies$ 18_n.

If one summarizes the argument without bringing in 18 then inevitably MT 1.n and MT 2.n are seen to depend on *all* previous instances of both results.

The classification theorem is largely contained in Part 1 of the Main Theorem but some additional information is needed to cover special cases, specifically:

PROPOSITION 14. *Let H be a primitive homogeneous directed graph embedding $L(2)$ and I_n and omitting I_{n+1}, where $n \geq 2$. If $n = 2$ assume also that H is not isomorphic with $S(3)$. Then all the directed graphs in $\mathcal{A}_n$ embed into H.*

The only nontrivial claim in the conclusion of this proposition is that $[I_1, I_2]$ embeds in H, and this is exactly what fails to happen in $S(3)$. The point is that since H contains I_2 and $L(2)$ and the relation $\perp$ (strictly, its reflexive closure) is not an equivalence relation, H must embed $I_1 + L(2)$. The proof of Proposition 14 will be given below.

We shall now check that Proposition 14 may be combined with the first part of the Main Theorem to yield the stated classification result. We suppose that H is primitive and homogeneous, omitting I_∞, and notice that there is then some n for which H embeds I_n and omits I_{n+1}. If $n = 1$ we are dealing with a homogeneous tournament, and we invoke Lachlan's classification, getting three examples given explicitly in the list, as well as the degenerate example I_1 included in the first group, and $T^\infty = \Gamma_1$. If $n > 1$ then according to the last proposition either H is isomorphic to one of I_n or $S(3)$, or else H embeds the three directed graphs of $\mathcal{A}_n$. Then the case 1.n of the Main Theorem applies to the class of finite directed graphs embedding in H, and says that this is the same as the class of finite directed graphs embedding in Γ_n. By Fraïssé's correspondence between amalgamation classes and homogeneous structures, $H \simeq \Gamma_n$ in this case.

The proof of Proposition 14 presents no special difficulties, but we defer it for a moment. Instead we shall formulate the major steps in the proof of the Main Theorem as a sequence of distinct propositions and then describe the inductive structure of the argument in greater detail.

5.4. The major steps in the proof

In addition to Proposition 14 we shall require six similarly concrete statements, some of which have quite elaborate proofs, from which our Main Theorem follows expeditiously, as we shall show. It will be convenient to use the "consequence" terminology, according to which a finite directed graph C is a consequence of two sets $\mathcal{A}$, $\mathcal{B}$ of finite directed graphs if it is in every amalgamation class containing the first set and omitting all graphs in the second. Our notation for this will be $\mathcal{A}\&\neg\mathcal{B} \implies C$, or some very similar expression. There is of course no claim that the set of consequences of a given collection of finite directed graphs will itself be an amalgamation class, though this occurs in many concrete cases.

The next proposition will be proved in the course of dealing with Proposition 14.

PROPOSITION 15. *For $n \geq 3$:*

$$\mathcal{A}_n \& \neg I_{n+1} \implies I_1 + \mathcal{A}_{n-1}$$

Here $I_1 + \mathcal{A}_{n-1}$ denotes $\{I_1 + A : A \in \mathcal{A}_{n-1}\}$.

The next five propositions will be proved in the following chapter.

PROPOSITION 16. *$\mathcal{A}_2 \& \neg I_3 \implies I_1 + C_3$, and any directed graph on four points which is an oriented form of the following configuration is also a consequence of $(\mathcal{A}_2 \& \neg I_3)$:*

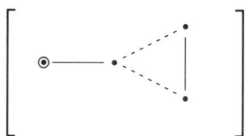

The convention here is that solid edges are to be oriented, dashed edges should be either oriented or omitted. The next proposition is the same as Proposition 15, but stated also for $n = 2$.

PROPOSITION 17. *$\mathcal{A}_n \& \neg I_{n+1} \implies (I_1 + \mathcal{A}_{n-1})$ for $n \geq 2$.*

PROPOSITION 18. *Let $A \not\simeq I_{n+1}$ be a one point extension of I_n. Then $\mathcal{A}_n \implies A$.*

PROPOSITION 19. *Let $\mathbb{H}$ be a homogeneous 2-directed graph satisfying:*
(1) *H_1 omits I_∞;*
(2) *$H_2 \simeq T^\infty$;*
(3) *$\mathbb{H}$ realizes (i.e., embeds) all 1-types over $L(2)$.*

Then for $y \in H_2$ and p a cross type, the 2-directed graph $({}^p y, y')$ has properties $(1-3)$.

PROPOSITION 20. *Let $\mathbb{H}$ be a homogeneous 2-directed graph satisfying:*
(1) *H_1 omits I_∞;*
(2) *$H_2 \simeq \Gamma_n$;*
(3) *$\mathbb{H}$ realizes all 1-types over $L(2)$ and all 1-types over I_n.*

Then for $y \in H_2$ and p a cross type, the 2-directed graph $({}^p y, y^\perp)$ has properties $(1-3)$ with n replaced by $n-1$.

In both propositions our standing hypothesis that there are at most three cross types is very much alive. These last two propositions are very similar in spirit, with the former useful at the outset while the latter allows a use of induction.

We now comment briefly on the logical dependencies among the various results. Proposition 14 will itself be proved by a self-contained inductive argument, which uses the classification of the homogeneous tournaments. In the course of this argument Proposition 15 will also be proved. Proposition 16 summarizes the results of some very specific amalgamation arguments. Proposition 17 again uses the classification of the homogeneous tournaments quite explicitly, as well as detailed amalgamations. As we indicated earlier, the proof of Proposition 18 is inductive, and the induction is not self-contained, but needs to be woven into the proof of Main Theorem, after the proof of MT 2.$(n-2)$ and before the proof of MT 1.n (the natural point would be just before MT 1.n). In the inductive step we use both a reduction to the case $n-1$ and instances of the Main Theorem, namely MT 1.$(n-2)$ and 2.$(n-2)$. The consequences of this have been sketched earlier.

The last two propositions are proved fairly directly by amalgamation arguments. The proof of Proposition 19 uses Proposition 15.

With the exception of Proposition 18, none of these propositions requires anything out of the ordinary, nor are the proofs particularly long. Furthermore, for the most part the completion of the inductive proofs of the two parts of the Main Theorem flows smoothly out of the stated propositions. There are two exceptions to this. In the first place, we have not been able to get a clean proof of the instance 2.2 of the Main Theorem, and consequently we shall spend a great deal of time on that one issue, and prove a large number of very minor results on the way. All of that material will be found in Chapter 6. The other exception is Proposition 18, where we make strong use of our global inductive hypothesis.

5.5. Proof of the Main Theorem, Part 1

We are going to show how the proof of the Main Theorem can be given on the basis of the propositions stated above. This section is devoted to the first part of the Main Theorem. For the whole of the present section the parameter n is fixed and the following are assumed known if $n > 1$:

(1) Propositions 16 and 17;
(2) Case n of Proposition 18;
(3) MT 1.$(n-1)$, MT 2.$(n-1)$.

We shall prove MT 1.n. For $n=1$ we simply refer to the classification of the homogeneous tournaments. We now assume $n>1$.

DEFINITION 5.7. Let $\mathcal{A}$ be an amalgamation class of finite directed graphs.
1. $\mathcal{A}^*$ is defined as the class of directed graphs $A \in \mathcal{A}$ such that:
    For any linear extension $B = A \cup L$ of A, if B omits I_{n+1} then $B \in \mathcal{A}$.
2. $\mathcal{A}^+$ is defined as the class of directed graphs $A \in \mathcal{A}$ such that:
    For any linear tournament L, and for every extension B of the wreath product $L[A]$ by a single vertex, if B omits I_{n+1} then B is in $\mathcal{A}$.

To say that $A \cup L$ is a linear extension of A means simply that L is a linear tournament. Observe that $\mathcal{A}^*$ will also be an amalgamation class, by a standard sort of argument based on the possibility of combining several linear tournaments into one large linear tournament. $\mathcal{A}^+$ has no obvious formal properties, but using this notation we can formulate two critical technical results.

LEMMA 5.8. *Let $\mathcal{A}$ be an amalgamation class of finite directed graphs containing the class $\mathcal{A}_n$ and omitting I_{n+1}. Then $\mathcal{A}_n \subseteq \mathcal{A}^+$.*

LEMMA 5.9. *Let $\mathcal{A}$ be an amalgamation class of finite directed graphs. Then $\mathcal{A}^+ \subseteq \mathcal{A}^*$.*

We shall now show that MT 1.n is contained in these two results. Using the correspondence between homogeneous structures and amalgamation classes, it suffices to show that if A is any finite directed graph omitting I_{n+1} and $\mathcal{A}$ is any amalgamation class containing $\mathcal{A}_n$ and omitting I_{n+1}, then $A \in \mathcal{A}$. We argue by induction on $|A|$.

The base case is $|A|=0$ (or $|A|=1$ if one prefers). For the inductive step, let $A = A_0 \dot\cup \{a\}$ where $|A_0| = |A|-1$, and let $\mathcal{A}$ be any amalgamation class containing $\mathcal{A}_n$ and omitting I_{n+1}. By Lemmas 5.8, 5.9, $\mathcal{A}^*$ will be an amalgamation class containing $\mathcal{A}_n$ and omitting I_{n+1}, so by induction A_0 belongs to $\mathcal{A}^*$. As A is a linear extension of A_0 (since I_1 is linear), it follows from the definition of $\mathcal{A}^*$ that A is in $\mathcal{A}$, as claimed.

This way of organizing an induction has its roots in [61], [57].

The proof of Lemma 5.9 is a standard application of Lachlan's Ramsey argument, which provides the main combinatorial trick in this area. Any linear extension of A can be forced to appear by amalgamating sufficiently many one-point extensions of $L[A]$, where L is a long linear tournament. If the desired linear extension of A omits I_{n+1} then so do the associated one-point extensions of $L[A]$, if they are chosen sensibly. This last point is critical, and represents the principal difficulty which may arise in doing such an argument. For more detail, see the proof of Proposition 4.

We still need to prove Lemma 5.8, and here something interesting occurs. Recall that $\mathcal{A}_n$ consists of $I_1 + L(2)$, $[I_1, I_2]$, and I_n. If A is one of these three directed graphs, and $B = L[A] \cup \{x\}$ with L a linear tournament, we want to show that B belongs to a certain amalgamation class, assuming that B omits I_{n+1}. We first make a structural observation.

LEMMA 5.10. *Let A be one of the directed graphs $I_1 + L(2)$, $[I_1, I_2]$, or I_n, and let $B = L[A] \cup \{x\}$ with L linear. If B omits I_{n+1} then there is a partition of the vertices of B into n induced tournaments, with at most one of them nonlinear. (This does not mean that all the edges lie within a set in the partition.)*

PROOF. It suffices to partition each of the sets $(\{a\} \times A) \cup \{x\}$ (for $a \in L$) into n linear tournaments, and to reassemble the pieces subsequently. The piece which contains x may end up being nonlinear. Now if $|A \cup \{x\}| \leq n+1$ then as the directed graph $A \cup \{x\}$ is not I_{n+1}, it has a partition into n linear tournaments. If $|A \cup \{x\}| > n+1$ then the only possibility is $|A| = 3$, hence $n = 2$. But any directed graph of order 4 omitting I_3 can be partitioned into two tournaments, at least one of which will be linear. $\square$

In view of this result, Lemma 5.8 is contained in the following result.

LEMMA 5.11. *Let $\mathcal{A}$ be an amalgamation class of finite directed graphs which contains $\mathcal{A}_n$ and omits I_{n+1}, and let A be a finite directed graph which can be partitioned into at most n tournaments in such a way that at least one of them is linear. Then $A \in \mathcal{A}$.*

PROOF. We may write A as the union of a linear tournament L and a directed graph A_0 which is the union of $n-1$ tournaments, and hence omits I_n. Associated with $\mathcal{A}$ there is a homogeneous directed graph H with the property that embeddability in H and membership in $\mathcal{A}$ are equivalent, for finite directed graphs. So we wish to embed A in H. Fix $x \in H$. We are going to find an embedding of A into H in such a way that $L \leq x'$ and $A_0 \leq x^\perp$. So consider the 2-directed graph $\mathbb{H} = (x', x^\perp)$. The idea is to apply MT 2.(n-1) to $\mathbb{H}$. This will certainly suffice. Accordingly we need only verify that MT 2.$(n-1)$ applies to $\mathbb{H}$, or explicitly:

(1) $x^\perp \simeq \Gamma_{n-1}$;
(2) $\mathbb{H}$ realizes all 1-types over $L(2)$, and all 1-types over I_{n-1}.

For (1) we apply the characterization given by MT 1.$(n-1)$, that is we claim that $\mathcal{A}_{n-1}$ embeds in $x^\perp$. As $\mathcal{A}$ contains $\mathcal{A}_n$ it suffices to show that $\mathcal{A}_n \implies (I_1 + \mathcal{A}_{n-1})$. But this is Proposition 17. Similarly the second claim may be seen to be contained in Propositions 16, 18. Here we first use Proposition 17 to see that for any $I \simeq I_{n-2}$ embedded in H, $I^\perp$ embeds $\mathcal{A}_2$ and not I_3, so the conclusion of Proposition 16 applies to $I^\perp$, and *a fortiori* to H. This provides the necessary 1-types over $L(2)$ in $\mathbb{H}$. $\square$

As we have indicated, this completes the proof of Lemma 5.8, and hence finishes the induction step in the proof of the Main Theorem, Part 1.

5.6. Proof of the Main Theorem, part 2, $n > 2$

This section is devoted to the second part of the Main Theorem. For the whole of the present section the parameter $n > 2$ is fixed and the following are assumed known:

(1) Proposition 20;
(2) MT 1.n, MT 2.$(n-1)$.

We shall derive MT 2.n here. We shall deal with the case $n = 1$ in the next section, but we defer the proof in the case $n = 2$ to Chapter 6, as it is rather elaborate.

So we are given a homogeneous 2-directed graph $\mathbb{H}$ and we assume:

(3) H_1 omits I_∞; $H_2 \simeq \Gamma_n$.
(4) All 1-types over $L(2)$ and all 1-types over I_n embed in $\mathbb{H}$

We have to embed an arbitrary 2-directed graph of the form (L, A) into $\mathbb{H}$, where L is linear and A omits I_{n+1}. In addition the cross types in all 2-directed graphs will be assumed be cross types of $\mathbb{H}$, without further mention.

Let $\mathcal{B}$ be the class of finite directed graphs A such that every finite 2-directed graph of the form (L, A) with L a linear tournament embeds in $\mathbb{H}$. MT 2.n then says that $\mathcal{B}$ contains all finite directed graphs omitting I_{n+1}. $\mathcal{B}$ is easily seen to be an amalgamation class, so MT 1.n applies. Accordingly it suffices to prove the following:

LEMMA 5.12. $\mathcal{A}_n \subseteq \mathcal{B}$.

PROOF. We first show that $\mathcal{A}_{n-1} \subseteq \mathcal{B}$. As $n > 2$ this will show in particular that $I_1 + L(2)$ and $[I_1, I_2]$ are in $\mathcal{B}$.

So fix $y \in H_2$, a cross type p, and let $\mathbb{H}^{(1)} = ({}^p y, y^\perp)$. Applying Proposition 20, MT 2.$(n-1)$ applies to $\mathbb{H}^{(1)}$, so for $A \in \mathcal{A}_{n-1}$, any linear extension (L, A) of A will embed in $\mathbb{H}^{(1)}$, and a fortiori in $\mathbb{H}$.

We must also check that $I_n \in \mathcal{B}$. By Lachlan's Ramsey argument, it suffices to embed 1-types $\mathbb{A}$ over $n \cdot L$ into $\mathbb{H}$, for L an arbitrary linear tournament, that is:

$$|A_1| = 1; \quad A_2 = n \cdot L.$$

In the Ramsey argument itself, $n \cdot L$ simply plays the role of a directed graph containing several disjoint copies of I_n; this particular choice has been made with a view toward the following argument.

Let L_0 be one of the linear components of $n \cdot L$, let $p = \text{tp}(A_1/L_0)$, and define $\mathbb{H}^{(2)} = ({}^p L_0, L_0^\perp)$. We shall show:

(∗) $\mathbb{H}^{(2)}$ realizes all 1-types over I_{n-1}, as well as over $L(2)$.

Then MT 2.$(n-1)$ can be applied to $\mathbb{H}^{(2)}$, and $\text{tp}(A_1/A_2 - L_0)$ is realized in $\mathbb{H}^{(2)}$, that is $\mathbb{A}$ is realized in $\mathbb{H}$.

Now in terms of $\mathbb{H}$, our assertion (∗) states that we can embed 1-types $\mathbb{A}$ over $L + I_{n-1}$ or over $L + L(2)$ into $\mathbb{H}$. Now as $n \geq 3$, we can apply Proposition 20 and MT 2.$(n-1)$ to conclude that 1-types over $L + L(2)$ embed in $\mathbb{H}^{(1)}$, hence in $\mathbb{H}$. Suppose therefore that $\mathbb{A}$ is a 1-type over $L + I_{n-1}$. Let L_0 be one of the vertices occurring as an isolated vertex in $L + I_{n-1}$, and repeat the previous argument, using $\mathbb{H}^{(2)}$. This reduces our problem to the embedding of 1-types over $L_0 + I_{n-1}$ and over $L_0 + L(2)$ into $\mathbb{H}$; the former is possible by hypothesis, the latter by the argument just given. □

This completes the inductive portion of the proof of MT 2.n for $n \geq 3$.

5.7. Case 2.1 of the Main Theorem

We shall now prove Part 2 of the Main Theorem for $n = 1$, using the classification of the homogeneous tournaments. So we are given a homogeneous 2-directed graph $\mathbb{H}$ in which H_1 omits I_∞, $H_2 \simeq T^\infty$, and $\mathbb{H}$ embeds all 1-types over $L(2)$.

Let $\mathcal{A}$ be the class of all finite tournaments T such that every linear extension (L, T) of T embeds in $\mathbb{H}$. This is an amalgamation class, and our claim is that it contains all finite tournaments. By the classification of homogeneous tournaments and the Fraïssé correspondence with amalgamation classes, it suffices to prove that $[I_1, C_3]$ belongs to $\mathcal{A}$. On the other hand by applying Proposition 19 and replacing

ℍ by a derived 2-directed graph, it suffices to prove (under the same hypotheses on ℍ) that $C_3 \in \mathcal{A}$. There is a routine way to do this, in two steps, which we now run through.

LEMMA 5.13. *If* 𝔸 *is a finite 2-directed graph with both components linear, then* 𝔸 *embeds in* ℍ.

PROOF. By Lachlan's Ramsey argument we can take $|A_1| = 1$, and proceed by induction on $|A_2|$, using Proposition 19 to carry out the inductive step. □

LEMMA 5.14. $C_3 \in \mathcal{A}$.

PROOF. By Lachlan's Ramsey argument it suffices to embed 1-types 𝔸 over $L[C_3]$ into ℍ, with L linear. With 𝔸 fixed, let C be the last copy of C_3 in $L[C_3]$, $p = \text{tp}(A_1/C)$, and $\mathbb{H}^* = (^pC, {'C})$. We wish to show that $\mathbb{H}^*$ also realizes all 1-types over $L(2)$, since we can then conclude rapidly by induction on $|L|$. In other words, we need only check that ℍ realizes all 1-types over $[L(2), C_3]$. After two applications of Proposition 19 and a slight change of notation, we reduce the problem to the consideration of 1-types over C_3 itself. There is a standard amalgamation diagram for this problem:

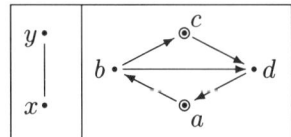

There are two constraints on this diagram: the factors omitting a or c should embed into ℍ, and after completion of the diagram by an oriented edge, one of the configurations x/abc, y/acd should be the desired 1-type. The choice of $\text{tp}(xd), \text{tp}(yb)$ is immaterial.

By the previous lemma the factor omitting a will embed in ℍ in any case. The factor omitting c is constructed by amalgamating $xyab$ with $yabd$. Here $xyab$ is afforded by the previous lemma, wile $yabd$ is manufactured by an amalgamation determining $\text{tp}(yb)$. □

This completes the verification of MT 2.1. We have seven propositions that still require proof, as well as the deferred case MT 2.2.

5.8. Propositions 14 and 15

We have two objectives in the present section. In the first place, we give the classification of the primitive homogeneous directed graphs omitting both I_∞ and $[I_1, I_2]$, which is the content of Proposition 14. We shall also show that $\mathcal{A}_n + \neg I_{n+1} \implies I_1 + \mathcal{A}_{n-1}$ for $n \geq 3$; the case $n = 2$ is treated later.

In the proof of the next lemma we shall become involved in some explicit amalgamation arguments of an elementary nature, which are quite simple once suitable strategies have been adopted. On the whole our policy is to give amalgamation diagrams explicitly, with the exception of those involving amalgamations of factors known to be available and with a unique plausible result. A result is implausible if it violates explicit hypotheses, notably omission of some I_{n+1} or some assumption associated with a division into cases, or if it would contain the configuration which is the ultimate goal of the construction (i.e., it violates an implicit working hypothesis).

5.8. PROPOSITIONS 14 AND 15

LEMMA 5.15. *Let H be a primitive homogeneous directed graph embedding I_2 and $L(2)$ and omitting I_3 and $[I_1, I_2]$. Then $H \simeq S(3)$.*

PROOF. Our hypotheses on H imply that $I_1 + L(2)$ embeds in H. We build up a little more detailed information about small configurations in H.

(1) $[\bullet \longrightarrow \bullet \longrightarrow \bullet]$ embeds in H

If this fails, then the domination relation "$x \longrightarrow y$" implies $x^\perp \supseteq y^\perp$, which is a nontrivial transitive relation. Since H is primitive, the relation defined by "$x^\perp = y^\perp$" must be equality. By homogeneity the relation of proper inclusion "$x^\perp \supset y^\perp$" is equivalent to a disjunction of 2-types, and as it is asymmetric it must coincide with the domination relation. Thus H is a partial order. By [**82**] there is a unique primitive homogeneous partial order embedding I_2 and L_2, and it embeds I_∞, which is a contradiction.

(2) $[\bullet \longrightarrow \bullet \longrightarrow \bullet \longrightarrow \bullet]$ embeds in H

This follows from a simple amalgamation based on (1), with a unique plausible result (an edge is inserted).

(3) H omits C_3

Assuming that C_3 embeds in H, we shall force $[I_1, I_2]$ into H by a suitable amalgamation. Our strategy is shown in the following diagram, where the dashed edge may be absent or oriented.

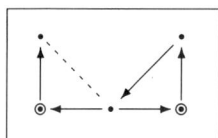

Any such diagram will suffice if its factors are present in H.

If $I_1 + C_3$ embeds in H then we can omit the dashed edge, as both factors then occur in H. On the other hand if H omits $I_1 + C_3$ then we can obtain the same diagram with the dashed edge present and oriented downward.

(4) H omits $[I_2, I_1]$

Assume $[I_2, I_1]$ embeds in H. The following amalgamation produces a contradiction (I_3 or C_3):

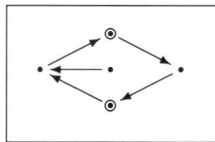

One factor is provided by (2). The other factor is forced into H by a similar construction, using $[I_2, I_1]$.

Now we can analyze the structure of H. We fix $x \in H$ and consider the 3-tournament $\mathbb{T} = (x', {}'x, x^\perp)$. The components are linear homogeneous tournaments, and it is easy to see they all embed $L(2)$, hence are isomorphic to $\mathbb{Q}$. As H omits $[I_1, I_2]$, $[I_2, I_1]$, and C_3, $\mathbb{T}$ is shuffled with two cross types in each 2-restriction. It follows that $\mathbb{T}$ is uniquely determined by this information, as can be checked very directly or read off from the more elaborate results of Chapters 2 and 3. Hence H is also uniquely determined. Since $S(3)$ has the required properties, $H \simeq S(3)$. □

This proves Proposition 14 for $n = 2$. The existence of the sporadic example $S(3)$ will lead us to view the case $n = 3$ as the base step of our inductive argument. This base step is the subject of the next three lemmas.

LEMMA 5.16. *Let H be a primitive homogeneous directed graph embedding $L(2)$ and I_3, and omitting I_4. Then H embeds $I_2 + L(2)$.*

PROOF. We suppose that H omits $I_2 + L(2)$. We build up a detailed knowledge of the small configurations in H until a contradiction emerges. In the first two steps, we show that $I_1 + P_3$ embeds in H; notice at the outset that if $I_1 + P_3$ is omitted then $I_1 + [I_1, I_2]$ and $I_1 + [I_2, I_1]$ are easily forced into H.

(1) P_3 embeds in H

Use: $[\cdot \longrightarrow \odot \longleftarrow \cdot \longleftarrow \odot \longrightarrow \cdot\,]$

(2) $I_1 + P_3$ embeds in H.

Use the following two amalgamation procedures, in order, adapting the orientations in the second one to the result of the first:

$[\odot \quad\quad \cdot \longrightarrow \odot \longrightarrow \cdot\,]$

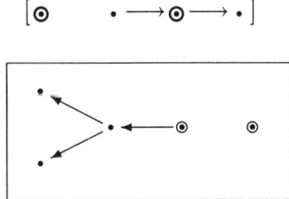

The first amalgamation will yield the desired result, since otherwise the second amalgamation yields $I_2 + L(2)$.

The next property is rather powerful.

(3) If $I \leq H$, $I \simeq I_3$, and $x \in H - I$, then x realizes either one or three types over the points of I.

Let $I = \{a, b, c\}$. Then c is definable from a, b and hence $\mathrm{tp}(xa), \mathrm{tp}(xb)$ determine $\mathrm{tp}(xc)$. Now our claim follows from (2).

(4) H omits the configuration:

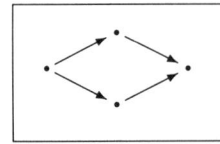

Otherwise the following amalgamation yields a contradiction to (3):

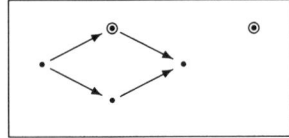

(5) H embeds $2 \cdot L(2)$

If this fails, then for $x \in H$ the relation $\perp$ defines a map ι from x' onto $x^\perp$, which is easily seen to be a bijection, as otherwise x' is imprimitive with connected components as equivalence classes (by (3)). Thus ι is an isomorphism, up to a permutation of the 2-types. But x' embeds I_3, and $x^\perp$ does not.

(6) $\mathbb{H}$ embeds $[\bullet \longleftarrow \bullet \longleftarrow \bullet \longrightarrow \bullet]$.

This is obtained from the following two amalgamations in succession, using the second one if the first one fails:
$$[\odot \quad \odot \longleftarrow \bullet \longrightarrow \bullet], \quad [\odot \longrightarrow \bullet \longleftarrow \bullet \longrightarrow \bullet \longleftarrow \odot].$$

(7) H embeds the two configurations following:

(7.1) 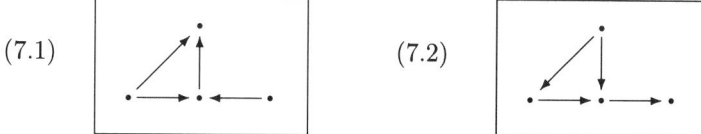 (7.2)

For (7.1) we use the amalgamation:

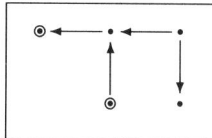

This has a unique completion, by (3), and that completion contains the desired configuration.

To build (7.2) we use:

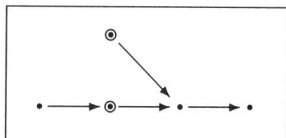

One factor is given by (2), and the other is constructed from:
$$[\bullet \longrightarrow \bullet \longrightarrow \odot \quad \odot \longrightarrow \bullet], \quad \text{using (3)}.$$

We are now within reach of a contradiction. Neither of the following amalgamation diagrams can be completed consistently:

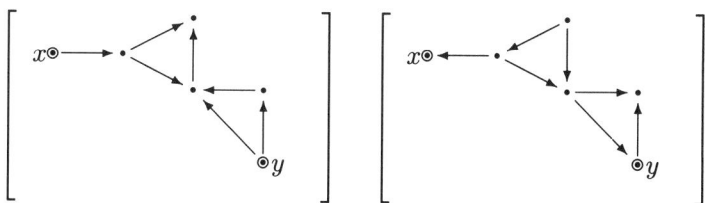

The two factors omitting y can both be embedded into H using

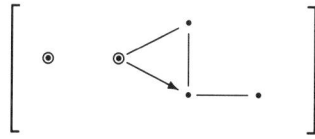

with a unique solution in both cases, by (3). So to complete the argument we need only produce one of the two factors omitting x, namely:

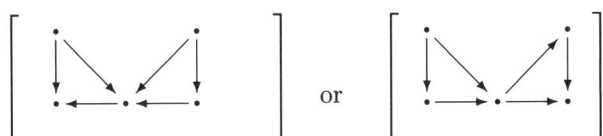

One of these will be forced by the amalgamation:

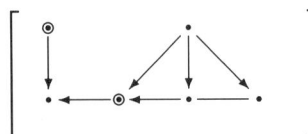

We have to examine the two factors involved. One is of the form $L(2) + L(3)$. If this is omitted by H then for $x \longrightarrow y$ in H, $(xy)^\perp$ will be isomorphic with C_3, and as y varies over x' this will yield too many 2-types in x'.

The second factor will be obtained by an amalgamation in which the orientation of the undetermined edge will be chosen. This amalgamation has one factor of the form (7.2), and the other factor is given by (6). □

LEMMA 5.17. *If H is a primitive homogeneous directed graph embedding I_3 and omitting I_4, then for $x \in H$ $x^\perp$ is also primitive.*

PROOF. By the previous lemma the only other possibility would be that $x^\perp \simeq 2 \cdot T$ for some homogeneous tournament T with $|T| > 1$. If this occurs, fix x and let T_1, T_2 be the two components of $x^\perp$. For $y \in T_1$, $y^\perp$ contains one component of $x^\perp$ and is disjoint from the other one. By symmetry T_2 is a component of both $x^\perp$ and $y^\perp$ for all $y \in T_1$. Now choose $y_1, y_2 \in T_1$ and $z \in T_2$. Then $y_1^\perp, y_2^\perp$ have the same component containing z. Since $\mathrm{tp}(xy_1y_2) = \mathrm{tp}(zy_1y_2)$, $y_1^\perp$ and $y_2^\perp$ must also have the same component containing x, that is $y_1^\perp = y_2^\perp$, and this is a nontrivial 0-definable equivalence relation on H. □

LEMMA 5.18. *If H is a primitive homogeneous directed graph embedding I_3 and omitting I_4, then H embeds $I_1 + [I_1, I_2]$.*

PROOF. According to Lemmas 5.15 and 5.17 the only alternative would be that

(∗) $x^\perp \simeq S(3)$ for $x \in H$.

In this case we show first: H omits $L(2) + P_3$

Assuming the contrary, we can form the amalgamation:

[• ⟵ • ⟵ ⊙   ⊙ ⟶ • ⟶ •]

which would produce $I_1 + [I_1, I_2]$.

Now we use the geometry of $S(3)$. For $y \in x'$ or $'x$, $y^\perp \cap x^\perp$ omits P_3, and hence $y^\perp$ is not dense in $x^\perp$. It follows easily that in these cases $(xy)^\perp$ is a cut in $x^\perp$ corresponding to a point in the dedekind completion. Thus x' and $'x$ have quotients isomorphic to $S(3)$, and therefore in fact $x' \simeq \, 'x \simeq S(3)$.

Fix $y \in x^\perp$. Then $y^\perp \simeq S(3)$ meets $'x$, $'x$, and $x^\perp$ in three cuts A_1, A_2, A_3. For $z \in A_1$, $z^\perp$ meets A_3 (since H embeds $I_2 + L(2)$) and contains y, so $z^\perp$ is not a cut in $x^\perp$, and this is a contradiction. □

The next two lemmas will cover the general case of Proposition 14 in the strong form needed to keep the induction on n going.

LEMMA 5.19. *Let H be a primitive homogeneous directed graph which embeds I_n and $L(2)$ and omits I_{n+1}, with $n \geq 3$. Then H embeds $I_2 + L(2)$.*

PROOF. We proceed by induction on n. Suppose on the contrary that for $x \in H$, $\perp$ is an equivalence relation on $x^\perp$, with classes of size $n-1$. We have treated the case $n=3$, so we assume $n>3$. Then we find that $x^\perp \simeq T_0[I_{n-1}]$ for some homogeneous tournament T_0. Here we are using the classification of the homogeneous directed graphs in the imprimitive case. As H is primitive, $|T_0| > 1$.

We shall get a contradiction by amalgamating $[I_1, I_n]$ with $I_1 + [I_1, I_{n-1}]$ over $[I_1, I_{n-1}]$. An edge must be inserted and this gives a copy of $I_2 + L(2)$. Now $[I_1, I_{n-1}]$ embeds in $T_0[I_{n-1}]$, so the second factor in our amalgamation is available in H. For the rest of the analysis we may therefore suppose:

(1) H omits $[I_1, I_n]$.

Then for $x \in H$, x' omits I_n and embeds I_{n-1}. We claim next:

(2) For $x \in H$, $x' \simeq T_1[I_{n-1}]$ for some homogeneous tournament T_1

We may suppose that $L(2)$ embeds in x'. Then induction applies, and either x' is primitive and embeds $I_2 + L(2)$, in which case we are done, or x' is imprimitive. As $n-1 \geq 3$, the classification in the imprimitive case shows that x' is a wreath product, either of the form $(n-1) \cdot T_1$ with $|T_1| > 1$, in which case we have an embedding of $I_2 + L(2)$ into H, or of the stated form $T_1[I_{n-1}]$, where T_1 is a homogeneous tournament.

The dual argument yields $'x \simeq T_2[I_{n-1}]$ for some homogeneous tournament T_2.

For $C \simeq I_{n-1}$, $C^\perp$ contains no edges and accordingly $|C^\perp| = 1$. The map $C \mapsto C^\perp$ carries the $\perp$-classes of x' onto the realizations of some nontrivial type over x. If this map is constant then its image is an x-definable element, and it follows rapidly that H is imprimitive. So we have an x-definable bijection between $x'/\perp$ and one of the sets $'x, x', x^\perp$, which must be an isomorphism up to a permutation of the types. However the domain is a tournament and the image is a nontrivial wreath product. □

LEMMA 5.20. *If the primitive homogeneous directed graph H embeds I_n and $L(2)$ and omits I_{n+1}, with $n \geq 3$, then H embeds $I_1 + [I_1, I_2]$.*

PROOF. We may take $n > 3$, and we suppose that H omits $I_1 + [I_1, I_2]$. For $x \in H$, if $x^\perp$ is primitive we conclude by induction. Assume therefore that $x^\perp$ is imprimitive. By the previous lemma we find $x^\perp \simeq (n-1) \cdot T_0$ for some homogeneous tournament T_0 with $|T_0| > 1$.

For $a \in x^\perp$, $a^\perp$ consists of $n-2$ components which contain components of $x^\perp$, and a copy of T_0 that contains x. If y is chosen to lie in this copy of T_0 then $y^\perp$ contains the same $n-2$ components of $x^\perp$, but does not contain $x^\perp$, since H does not carry any nontrivial transitive relation.

Let b lie in a component of $x^\perp$ contained in $y^\perp$. Then tp$(xay) = $ tp(xby), but in $x^\perp$ we have: $a/\perp \not\subseteq y^\perp$, $b/\perp \subseteq y^\perp$. This is a contradiction. □

In particular we have now proved Proposition 14 in a strong form which also yields Proposition 15:

For $n \geq 3$:

$$\mathcal{A}_n \& \neg I_{n+1} \implies I_1 + \mathcal{A}_{n-1}$$

It is important to rephrase this as follows: If a finite directed graph A is known to belong to every amalgamation class containing $\mathcal{A}_{n-1}$ to which I_n does not belong, then $I_1 + A$ belongs to every amalgamation class containing $\mathcal{A}_n$ to which I_{n+1} does not belong. This version may be more transparent if rephrased in terms of homogeneous directed graphs.

CHAPTER 6

Propositions 16 to 20 and MT 2.2

In the present chapter we prove the five technical propositions on which the arguments of the last chapter were based. As explained there, one of them, Proposition 18, cannot be proved independently of the main theorems, but is proved within the same inductive framework as the main theorems. We deal with this by simply recording the relevant inductive hypothesis before proving Proposition 18, in accordance with the inductive structure of the proof as a whole which was laid out in the preceding chapter.

6.1. Proposition 16: simple configurations

Proposition 16 stated that an amalgamation class $\mathcal{A}$ containing $\mathcal{A}_2$ and omitting I_3 must contain $I_1 + C_3$ as well as every directed graph of the form:

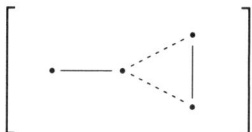

where the solid edge is to be oriented and the dashed edges may be oriented or omitted. We shall now prove this.

PROOF. We use very concrete amalgamation arguments. We shall show that all of the following configurations are in $\mathcal{A}$:

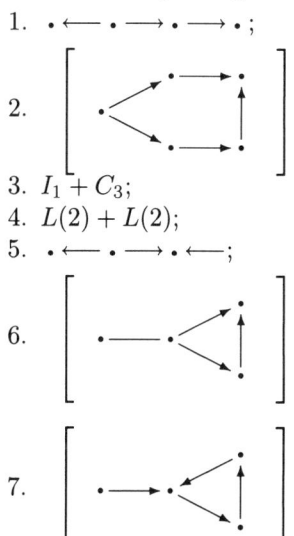

1. $\cdot \longleftarrow \cdot \longrightarrow \cdot \longrightarrow \cdot$;
2.
3. $I_1 + C_3$;
4. $L(2) + L(2)$;
5. $\cdot \longleftarrow \cdot \longrightarrow \cdot \longleftarrow \cdot$;
6.
7.

8. $\begin{bmatrix} \cdot \longrightarrow \cdot \triangleleft \cdot \end{bmatrix}$

We first consider the following series of amalgamations:
$[\odot \quad \odot \longleftarrow \cdot \longrightarrow \cdot]$, $[\odot \longrightarrow \cdot \longleftarrow \cdot \longrightarrow \cdot \longleftarrow \odot]$,
$[\odot \longleftarrow \cdot \longleftarrow \cdot \longrightarrow \cdot \longrightarrow \odot]$, $[\odot \longrightarrow \cdot \longrightarrow \cdot \longrightarrow \cdot \longrightarrow \odot]$.

The first amalgamation produces either (1) or (5). In the presence of (5) the second amalgamation will produce (1). So $\mathcal{A}$ contains the configuration (1). Now the third amalgamation yields (2), after which the fourth yields some form of:

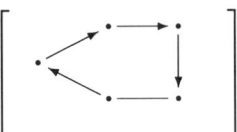

This allows us to form:

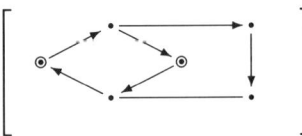

which yields (3) and (4), as well as either (7) or its dual.

Now we obtain (5). Suppose first that (7) is in $\mathcal{A}$. We then form the following two amalgamations:

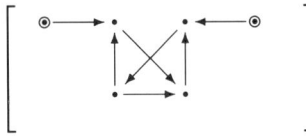

which either yield (5), or else produce the factors for a further amalgamation, which produces (5):

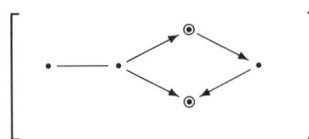

If (7) is not in $\mathcal{A}$, then its dual is, and as (5) is self-dual the same argument works.

For (6) we may use: $\begin{bmatrix} \cdot \longrightarrow \cdot \diamond \cdot \end{bmatrix}$

For (7) and (8) we consider the two configurations:

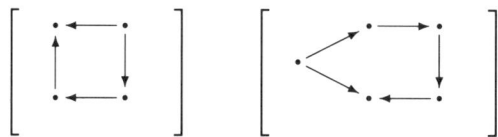

If one of these belongs to $\mathcal{A}$ then we derive (7,8) from one of the following two diagrams:

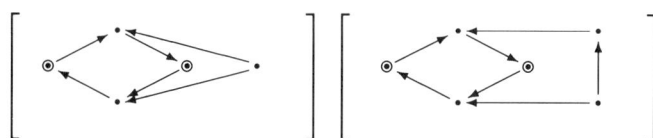

If we assume that neither of these configurations is in $\mathcal{A}$, then the following two amalgamations produce (7) and (8) respectively:

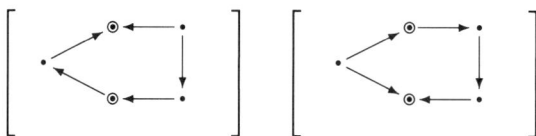

We have now taken care of all instances of Proposition 16, up to duality. □

6.2. Proposition 17: induction on n

Proposition 17 extends Proposition 15 by including the case $n = 2$, namely: there is no homogeneous directed graph H embedding $\mathcal{A}_2$ and omitting both I_3 and $I_1 + [I_1, C_3]$. We fix such a directed graph H throughout the present section, and we aim at a contradiction. We shall make use of Proposition 16.

Fix a point $x \in H$ and let $\mathbb{H}$ be the 2-directed graph $(x^\perp, x')$. We have assumed that $x^\perp$ is a nongeneric tournament. By Proposition 16 $x^\perp$ embeds C_3. If $x^\perp \simeq C_3$ then as H is primitive it is finite. This case can be eliminated fairly easily, but we prefer to refer to the classification of the finite homogeneous directed graphs [56]. So we may suppose that $H_1 = x^\perp \simeq S(2)$.

LEMMA 6.1. *The directed graphs in $\mathcal{A}_2$ embed in H_2.*

PROOF. By Proposition 16 $I_1 + L(2)$ embeds in H_2. We need to embed $[I_1, I_2]$ in H_2, or equivalently, $[L(2), I_2]$ in H. Our first try is:

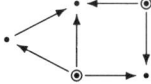

The insertion of an edge produces $[L(2), I_2]$. If the amalgam contains no new edge then H contains the factors of:

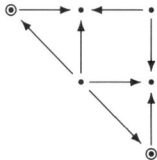

This then forces $[L(2), I_2]$ into H.

□

LEMMA 6.2. *Any finite 2-directed graph $\mathbb{A}$ whose components are both linear tournaments embeds in $\mathbb{H}$.*

PROOF. By Lachlan's Ramsey argument it suffices to consider the case in which A_1 is linear and A_2 is a singleton. It then suffices to show that for $a \in H_2$ and for any cross type p, $^p a$ is dense in H_1. If this fails then we have a map from H_1 into the dedekind completion of H_1, and as H_2 is primitive but is not a tournament, this yields a contradiction. □

Now we are ready to prove Proposition 17.

PROOF. Let $\mathbb{A} = (I_1, C_3)$ with $I_1 \perp C_3$. We are going to embed $\mathbb{A}$ in $\mathbb{H}$, and hence embed $I_1 + [I_1, C_3]$ in H. We use an amalgamation of the following form, whose factors we display separately:

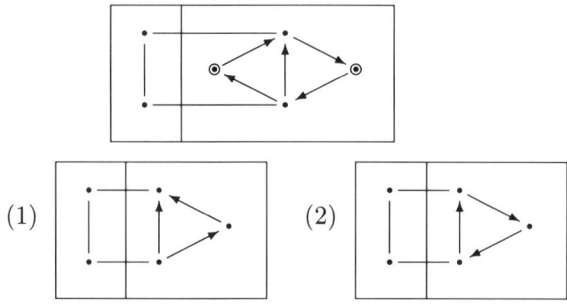

By the preceding lemma, any form of (1) is available in $\mathbb{H}$, so we need only find a factor of type (2) in $\mathbb{H}$. Since $\mathbb{H}$ omits any form of I_3, we can construct a form of (2) by an amalgamation which chooses the orientation of the edge in the first component, assuming we have factors of the form:

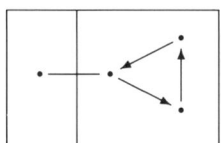

If we determine the cross type by an amalgamation, we either get a suitable form of this factor, or (I_1, C_3) itself. For this last amalgamation, one factor is covered by the previous lemma, while the other corresponds to an embedding of C_3 into H_2. Lemma 6.1 states that H_2 embeds the directed graphs of $\mathcal{A}_2$. By Proposition 16, C_3 embeds in H_2. □

6.3. Proposition 18: extending I_n

Proposition 18 states that any homogeneous directed graph H which embeds the directed graphs of $\mathcal{A}_n$ and omits I_{n+1} must embed every one-point extension of I_n except I_{n+1}.

Fix $n \geq 1$. The standing hypothesis for the present section is:

The instances $1.(n-1)$, $2.(n-1)$ of the Main Theorem are known.

We have explained previously that the proof of this proposition is not a self-contained induction, but is woven into the inductive proof of the Main Theorem.

We also take Proposition 17 as known. With these hypotheses, we prove Proposition 18.

PROOF. Let $A = \{a\} \cup I_n$ be the directed graph we wish to embed in H. Let $k = |a' \cap I_n|$, $l = |'a \cap I_n|$. If $n = 1$ there is nothing to prove, and if $n = 2$ then Lemma 16 applies. Accordingly we shall assume $n \geq 3$ throughout. In addition, if $k + l < n$ then $A = I_1 + A^-$ with A^- omitting I_n, and Proposition 17 together with MT $1.(n-1)$ covers this case. So we may also suppose that $k + l = n$.

Notice that $\mathbb{H}$ contains $\Gamma(n-1)$, by Proposition 17 and MT $1.(n-1)$.

Case 1. $k = n$.

Fix $a \in H$. If H omits A then a' omits I_n, and by MT $1.(n-1)$, $a' \simeq \Gamma_{n-1}$. Let $\mathbb{H} = (a^\perp, a')$. Then H_1 omits I_∞, $H_2 \simeq \Gamma_{n-1}$, and $\mathbb{H}$ realizes all 1-types over I_{n-1} and $L(2)$ since the corresponding configurations in H either omit I_n, or are of the form $I_1 + K$ with K omitting I_n. By MT $2.(n-1)$, $\mathbb{H}$ realizes all forms of $(L(2), I_{n-1})$, and all one point extensions of I_n other than I_{n+1} occur as the underlying directed graphs of such configurations.

The case $l = n$ is of course treated similarly.

In the remaining cases we shall consider the particular directed graph B with vertex set $\{x_1, x_2\} \dot\cup I_n$, and with edges $x_1 \longrightarrow I_n \longrightarrow x_2$, $x_1 \longrightarrow x_2$.

Case 2. H omits B.

Fix $x_1 \longrightarrow x_2$ in H, and let $\mathbb{H} = ((x_1 x_2)^\perp, x_1' \cap 'x_2)$. Then H_1 omits I_∞, and by hypothesis H_2 omits I_n, so by MT $1.(n-1)$, $H_2 \simeq \Gamma_{n-1}$. Furthermore $\mathbb{H}$ realizes every 1-type over I_{n-1} and every 1-type over $L(2)$, since in H these either omit I_n, or are of the form $I_1 + K$ with K omitting I_n. Thus $\mathbb{H}$ embeds every form of $(L(2), I_{n-1})$, and H embeds every one point extension of I_n except I_{n+1}.

Case 3. H embeds B, and $k, l \geq 1$.

Write $B = \{x_1, x_2\} \dot\cup \{a_1, y\} \dot\cup Y$ with $a_1 y Y \simeq I_n$. Let $C = a_1 Y X_1 X_2$ with:

(1) $x_i \in X_i \simeq I_{n-1}$ ($i = 1, 2$);
(2) $X_1 \longrightarrow Y \longrightarrow X_2$, $X_1 \longrightarrow X_2$;
(3) $|a_1' \cap X_1| = k-1$, $|'a_1 \cap X_1| = l$, $|a_1' \cap X_2| = k$, $|'a_1 \cap X_2| = l-1$;
(4) $a_1 Y \simeq I_{n-1}$, $a_1 \longleftarrow x_1$, $a_1 \longrightarrow x_2$.

Then B, C agree on their common part $a_1 x_1 x_2 Y$. Observe that any pair of unlinked vertices of C belong to $a_1 Y, X_1$, or X_2, and in particular C omits I_n, hence embeds in H. In an amalgam $B \cup C$ in H of B and C over their common part, no identifications are possible between y and an element of $X_1 \cup X_2$. Furthermore we notice:

(5) $B \cup C - \{a_1\}$ omits I_n

$B \cup C - \{a_1\}$ is the union of $(X_1 - \{x_1\}) \dot\cup (X_2 - \{x_2\}) \dot\cup Y$, which omits I_{n-1}, and $\{x_1, x_2, y\}$, which omits I_2, so (5) follows.

By (5), $(B \cup C - \{a_1\}) + \{a_2\}$ embeds in H (with a_2 a new vertex), and can therefore be amalgamated with $B \cup C$ in H over their common part. In this amalgamation a_1 cannot be identified with a_2, nor can $a_1 \perp a_2$ hold, since this would force $a_1 a_2 y Y \simeq I_{n+1}$, so we have the possibilities:

$$a_1 \longrightarrow a_2; \quad a_1 X_1 a_2 \simeq A;$$

$$a_1 \longleftarrow a_1; \quad a_1 X_2 a_2 \simeq A.$$

Thus in any case A embeds in H. □

6.4. Proposition 19: $(^p y, y')$

In the present section we take Proposition 17 as known.

Proposition 19 states that if $\mathbb{H}$ is a homogeneous 2-directed graph whose first component omits I_∞, whose second component is T^∞, and which realizes all 1-types over $L(2)$, then for any $y \in H_2$ and any cross type p, the derived 2-directed graph $\mathbb{H}^* = (^p y, y')$ also embeds all 1-types over $L(2)$.

PROOF. We fix $\mathbb{H}, y, p$ as described. Our claim is that $\mathbb{H}$ embeds all 1-types over $L(3)$. We fix a 1-type $\mathbb{A}$ over $L(3)$, $\mathbb{A} = (\{x\}, \{a, b, c\})$, $a \longrightarrow b \longrightarrow c$ linear, and we let p, q, r be the type of x over a, b, c respectively. We shall use various concrete amalgamations.

Case 1. The types p, q, r are distinct.

In this case we use a cross amalgamation:

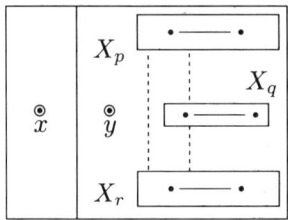

arranged so that if $\text{tp}(xy) = t$ then xyX_t will be isomorphic with $\mathbb{A}$. The only point that requires attention is the possibility of identifications occurring when the factor omitting y is formed, and the only disagreeable case is the one in which the two realizations of q are identified, since y should dominate one and be dominated by the other. To avoid this, we first form the structure $(\{x\}, X_p \cup X_r)$ by the amalgamation:

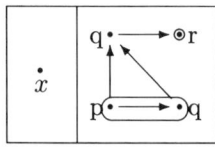

In this diagram one factor is a 1-type over $L(2)$. To see that the other factor is available one looks at the 2-tournament (x^p, x^q) which has only two cross types and is therefore not diagonal.

Case 2. $q = p$ or $q = r$.

This case is treated by varying the parameter c or a respectively. For example for $c \in H_2$, a c-definable bijection between two types over c in H_1 and H_2 respectively gives rise to a (c_1, c_2)-definable bijection between two types over c_1, c_2 in H_2, which by inspection must be the identity. So all these bijections would cohere and be restrictions of a 0-definable bijection, which does not exist since all 1-types over $L(2)$ are realized in $\mathbb{H}$.

Case 3. $\quad p = r$.

If H_1 is imprimitive then it may be replaced by a single equivalence class without affecting the hypotheses. Thus we may take H_1 to be primitive. If H_1 is a tournament then the classification of homogeneous 2-tournaments yields the result. Accordingly we may assume that I_2 embeds in H_1.

Suppose that $H_1 \simeq S(3)$. Then for $a \in H_2$, ^pa will be dense in H_1, as otherwise the convex closure of ^pa in $S(3)$ is a cut in $S(3)$, or an element of $S(3)$, and T^∞ embeds into the dedekind completion of $S(3)$, a contradiction. Consequently $(^pa, a')$ is a 2-directed graph of type $(S(3), T^\infty)$ in which the types p, q are both realized. It then follows easily that for any cross type s, and any $b \in a'$, $^sb \cap {}^pa$ will be dense in ^pa, and that every 1-type over $L \subseteq a'$ with $L \simeq L(2)$ will be realized in ^pa. Thus our claim follows in this case.

We shall therefore assume that H_1 is primitive embedding I_2, and that $H_1 \not\simeq S(3)$. By Proposition 17, H_1 embeds $I_1 + [I_1, C_3]$. Thus for $x \in H_1$, the directed graphs $x^\perp, x'$, and $'x$ all embed C_3.

For $x \in H_1$, the 2-directed graph $\mathbb{H}^* = (x^p, x^q)$ is a homogeneous 2-tournament with two cross types, which yields the desired result unless:

(*) (x^p, x^q) is shuffled with components of type $\mathbb{Q}$.

We analyze this case further.

Suppose first that we have $x_1, x_2 \in H_1$ with $x_1^p \cap x_2^p = \emptyset$. If $x_1 \perp x_2$, then for $a \in H_2$ we find that $(^pa, a')$ is a 2-tournament. This 2-tournament is nondiagonal by Case 2, with $a' \simeq T^\infty$. From the classification of homogeneous 2-tournaments it follows that for $x \in {}^pa$ we have $x^p \simeq T^\infty$, contradicting (*). On the other had if $x_1 \longrightarrow x_2$ then for $a \in H_2$ we have $^pa \simeq I_m$ for some $m < \infty$ and then $\bigcup_{x \in {}^pa} x^p$ is a finite union of copies of $\mathbb{Q}$, and is a-definable in H_2, a contradiction.

Thus for $x_1, x_2 \in H_1$ we find that x_1^p and x_2^p meet. Fix a nontrivial 2-type s in H_1. If $x, y \in H_1$ with $\text{tp}(xy) = s$ and $(xy)^p$ is not dense in x^p, then the ordering on the dedekind completion of x^p will induce a quasiordering on x^s, with quotient $\mathbb{Q}$. Since x^s is a homogeneous directed graph embedding C_3, this is impossible: the edge relation must be contained in the induced equivalence relation, so the quotient is symmetric.

We have shown that for $x, y \in H_1$ the set $x^p \cap y^p$ is dense in x^p. Choose y so that y^q meets x^p. Then the desired 1-type over $L(3)$ embeds in $(\{y\}, x^p)$. □

6.5. Proposition 20: $(^py, y^\perp)$

Proposition 20 says that if $\mathbb{H}$ is a homogeneous 2-directed graph such that H_1 omits I_∞, $H_2 \simeq \Gamma_n$ with $n > 1$, and which embeds all 1-types over $L(2)$ and all 1-types over I_n, then for $y \in H_2$ and for p any cross type, the induced 2-directed graph $(^py, y^\perp)$ embeds all 1-types over $L(2)$.

PROOF. We employ the same sort of concrete amalgamations used in the previous section. In terms of $\mathbb{H}$ we are concerned with configurations of the form shown:

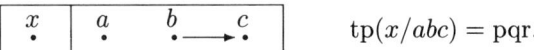 $\operatorname{tp}(x/abc) = \mathrm{pqr}.$

Let p, q, r denote the type of x over a, b, c respectively.

Case 1. The types p, q, r are distinct.

We use a cross amalgamation:

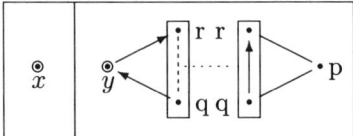

which is easily constructed if we have the factors shown:

(1) 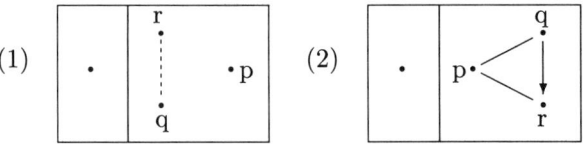 (2)

In factor (2) the two edges whose orientations have not been specified must be present. Otherwise, when $n = 2$, the factor omitting x in our main amalgamation would contain a copy of I_3 in its second component. The type of the optional edge in factor (1) is immaterial.

We can form a suitable version of (2) using:

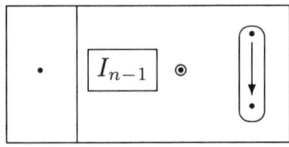

One factor is a 1-type over I_n and is therefore available in all possible forms. The other factor can easily be manufactured in some form.

Case 2. $q = r$.

Variation of the parameter a yields the result.

Case 3. $p = q$.

We shall represent the type r by an edge, and the type p by the absence of an edge. Accordingly we are concerned with the configuration:

We shall use the amalgamation:

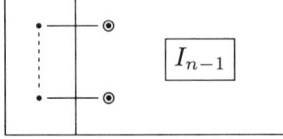

A suitable pair of factors is obtained from:

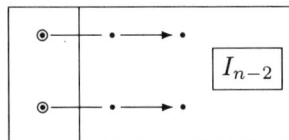

whose factors are:

From the following amalgamation we derive either this configuration or the 1-type we originally sought.

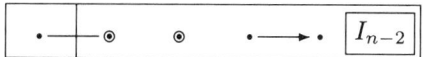

The factors for this are:

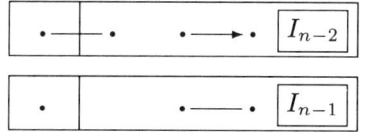

which are obtained by variation of parameters. □

6.6. Toward MT 2.2

The remainder of the present chapter concerns the instance MT 2.2 of our main theorem. So we consider a homogeneous 2-directed graph $\mathbb{H}$ satisfying:

(1)   The first component H_1 of $\mathbb{H}$ omits I_∞.

(2)   The second component H_2 of $\mathbb{H}$ is isomorphic with Γ_2.

(3) $\mathbb{H}$ embeds all 1-types over I_2 and all 1-types over $L(2)$.

Our claim is then that

(∗) $\mathbb{H}$ embeds all 2-directed graphs of the form (L, A) with A omitting I_3.

The standing hypothesis until the end of the chapter will be that MT 1.2 and MT 2.1 are known. We recall these statements:

(*MT*1.2) If an amalgamation class $\mathcal{A}$ contains $\mathcal{A}_2$ and omits I_3, then $\mathcal{A}$ consists of exactly the directed graphs omitting I_3.

(*MT*2.1) Let $\mathbb{H}$ be a homogeneous 2-directed graph whose first component H_1 omits I_∞, and whose second component H_2 is isomorphic with T_∞. Assume that $\mathbb{H}$ embeds all 1-types over $L(2)$ and that there are at least two cross types. Then $\mathbb{H}$ embeds all finite 2-directed graphs $\mathbb{A}$ for which A_1 is linear and A_2 is a tournament.

The argument can be viewed as taking place in three stages which involve four intermediate lemmas, given as Lemmas 6.3, 6.4, 6.5, and 6.6 below. Lemmas 6.5 and 6.6 are extremely special consequences of explicit amalgamation constructions, some of them quite lengthy. Based on these two very special lemmas, some type realization statements can be proved which clearly go in the general direction of MT 2.2. These are given as Lemmas 6.3 and 6.4. Finally it can be checked that MT 2.2 itself is a consequence of these two special cases.

Naturally we take this all in reverse order. We being by formulating two critical special cases of MT 2.2 from which we derive MT 2.2. In the next two sections we shall formulate Lemmas 6.5 and 6.6 and derive Lemmas 6.3 and 6.4 from them. In the last two sections we prove Lemmas 6.5 and 6.6. The two critical special cases of MT 2.2 are as follows.

LEMMA 6.3. *Under the hypotheses of MT 2.2, every 1-type over* $[I_1, I_2]$ *embeds in* $\mathbb{H}$.

LEMMA 6.4. *Under the hypotheses of MT 2.2, every 1-type over* $2 \cdot L(2)$ *embeds in* $\mathbb{H}$.

We now show how MT 2.2 follows from these two lemmas by the usual sorts of reductions. Accordingly we assume Lemmas 6.3 and 6.4 for the remainder of this section. First of all we derive much stronger forms of Lemmas 6.3 and 6.4 by the usual sorts of iterations and reductions:

PROPOSITION 21. *Under the hypotheses of MT 2.2, every 1-type over* $2 \cdot L$ *embeds in* $\mathbb{H}$, *for* L *a finite linear tournament*.

PROPOSITION 22. *Under the hypotheses of MT 2.2, for* L *a finite linear tournament, every 1-type over* $[I_1, I_2][L]$ *embeds in* $\mathbb{H}$.

PROPOSITION 23. *Under the hypotheses of MT 2.2, every 2-directed graph of the form* (L, A) *with* $A \in \mathcal{A}_2$ *embeds in* $\mathbb{H}$, *for* L *a finite linear tournament*.

This last Proposition is equivalent to MT 2.2, since we may consider the amalgamation class $\mathcal{B}$ of all tournaments A with the property that any 2-directed graph (L, A) with L a finite linear tournament embeds in every 2-directed graph satisfying the hypotheses of MT 2.2. The last proposition then shows that $\mathcal{B}$ contains $\mathcal{A}_2$, and then MT 2.1 shows that $\mathcal{B}$ contains every finite directed graph omitting I_3.

We now derive these three propositions. We deal first with Proposition 21. Let $L \leq H_2$ be a finite linear tournament. Let p be a 1-type over L, and $\mathbb{H}^* = (^pL, L^\perp)$. Our claim is that $\mathbb{H}^*$ embeds all 1-types over any finite linear tournament. By MT 2.1 it suffices to show that $\mathbb{H}^*$ embeds all 1-types over $L(2)$. In terms of the original 2-directed graph $\mathbb{H}$, this means that we have managed to replace $2 \cdot L$ by $L + L(2)$. A second application of the same argument reduces the problem to the case covered by Lemma 6.4.

We turn to Proposition 22. Note that $[I_1, I_2][L] = [L, 2 \cdot L]$. We argue by induction on $|L|$ that for any linear tournament L^* and any 2-directed graph $\mathbb{H}$ satisfying the hypotheses of MT 2.2, every 1-type over $[L, 2 \cdot L^*]$ will embed in $\mathbb{H}$. If $|L| = 0$ we have Proposition 21. If $|L| > 0$ let a be the first element of L, $L_0 = L - \{a\}$, so $[L, 2 \cdot L^*] = [\{a\}, [L_0, 2 \cdot L^*]]$. We pass to the derived 2-directed graph $\mathbb{H}^* = (^pa, a')$ for a suitable cross type p. It suffices to show that every 1-type over $[L_0, [L^*, L^*]]$ embeds into $\mathbb{H}^*$, and by induction this will follow if $\mathbb{H}^*$ satisfies the hypotheses of MT 2.2, in other words we must check that $\mathbb{H}^*$ embeds all 1-types over $L(2)$ and all 1-types over I_2. Equivalently we are claiming that $\mathbb{H}$ embeds all 1-types over $L(3)$ and all 1-types over $[I_1, I_2]$. This information is contained in Proposition 21 together with Lemma 6.3.

The last proposition concerns configurations of the form (L, A) with L a finite linear tournament and A one of the directed graphs $I_1 + L(2)$ or $[I_1, I_2]$. By Lachlan's Ramsey argument we may replace L by a singleton at the cost of replacing A by a graph embedding sufficiently many disjoint copies of A. Graphs of the form $2 \cdot L$ contain arbitrarily many copies of $I_1 + L(2)$, and graphs of the form $[I_1, I_2][L]$ contain arbitrarily many copies of $[I_1, I_2]$, so the necessary information is found in Propositions 21, 22.

6.7. Lemma 6.3

It remains only to prove Lemmas 6.3 and 6.4. At this point we need some very detailed information with particularly tedious proofs, given as Lemmas 6.5 and 6.6 below. The proofs of these lemmas will be given afterward. As the configurations under consideration in these lemmas will involve only two of the cross types present, we shall adopt the convention that one of these types, called p, is represented by a solid edge, and the other, q, is represented simply by the absence of an edge.

LEMMA 6.5. *Let $\mathbb{H}$ be a homogeneous 2-directed graph with H_1 omitting I_∞ and with $H_2 \simeq \Gamma_m$ for some m. Assume that $\mathbb{H}$ omits the following configurations:*

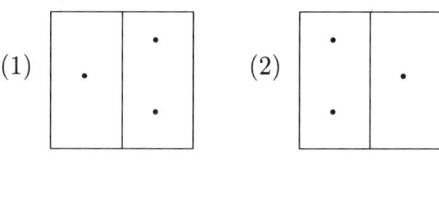

and embeds:

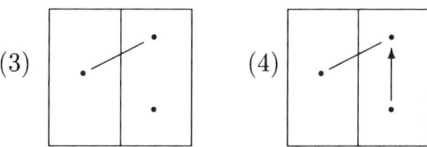

Then $\mathbb{H}$ also embeds the configurations shown:

LEMMA 6.6. *Let $\mathbb{H}$ be a homogeneous 2-directed graph with H_1 omitting I_∞ and with $H_2 \simeq \Gamma_2$. Assume that $\mathbb{H}$ omits the following configurations:*

and embeds:

Then $\mathbb{H}$ also embeds:

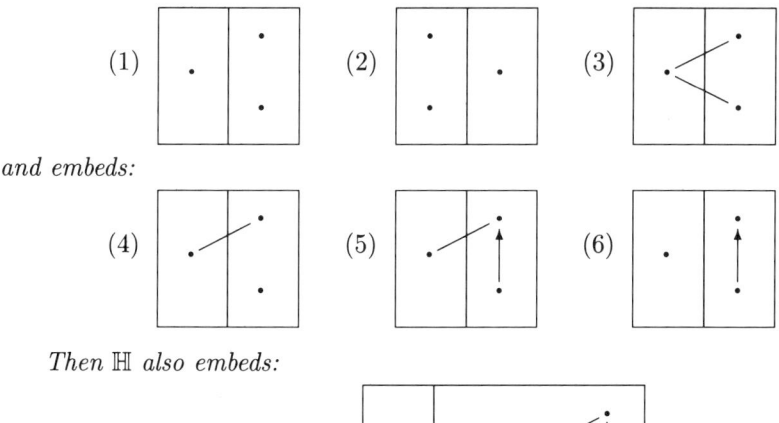

The proofs of these lemmas will be given at the end of the chapter. Now we take up the proof of Lemma 6.3. Frequently we shall be performing amalgamations involving 2-directed graphs in which only one vertex lies in the first component. In that case we shall generally exhibit only the second component, with some vertices labelled by the types they realize over the first component. As usual, type names lie in the range $p-r$ and vertex names in the ranges $a-c, x-z$. If only one vertex is circled, its type over the first component is to be determined by the amalgamation. In this case we speak of a "cross amalgamation".

PROOF. We note at the outset that Proposition 20 may be expressed as follows: any homogeneous 2-directed graph $\mathbb{H}$ satisfying the hypotheses of MT 2.2. will embed every 1-type over $I_1 + L_2$.

Now fix a homogeneous 2-directed graph $\mathbb{H}$ satisfying the hypotheses of MT 2.2, and let $\mathbb{A} = (\{a\}, [\{b\}, I_2])$. We wish to embed $\mathbb{A}$ in $\mathbb{H}$. Let p, q_1, q_2 be the types realized by a over b and the two vertices of I_2, respectively.

Case 1. $p = q_1$.

We use an amalgamation with two isomorphic factors; these are 2-directed graphs with a unique vertex in the first component, displayed in accordance with the convention introduced above:

$$[\underset{p}{\odot}\longrightarrow\underset{q_2}{\bullet}\quad\underset{?}{\bullet}\longleftarrow\underset{q_2}{\bullet}\quad\underset{q_2}{\bullet}\longrightarrow\underset{p}{\bullet}\longleftarrow\underset{p}{\odot}]$$

The necessary factor is constructed by an amalgamation in which the undetermined 2-type is chosen. One factor in this amalgamation lies wholly in the second component, and the other is afforded by Proposition 20 as reformulated above.

Case 2.   All three types p, q_1, q_2 are distinct.

We use a cross amalgamation of the form:

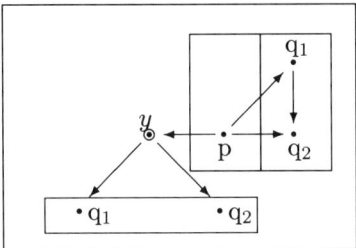

We only need to manufacture the factor omitting y, which is done by amalgamating the first two configurations shown.

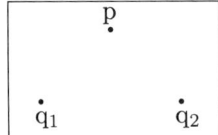 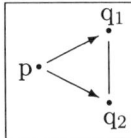 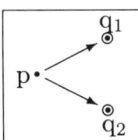

The first one is present by hypothesis, while the second is obtained by the amalgamation shown third.

Case 3. $q_1 = q_2 = q$.

We shall have to display both components of the 2-directed graphs used in our argument. We adopt the convention that the type p is represented by a solid edge, and the type q by the absence of an edge. Dashed edges within a component may be oriented, or absent.

Observe that for $x \in H_1$ and any 2-type r realized in $H_1 \times H_2$ we have $x^r \simeq \Gamma_2$. This follows from MT1.2 if we observe that x^r contains $\mathcal{A}(2)$ by hypothesis, Proposition 20, and (with regard to $[I_1, I_2]$) by a simple variation of parameters.

Our strategy is to form the following amalgamation.

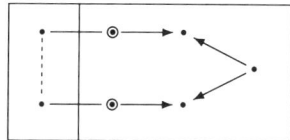

The factors here are both of the form:

If no edge is present two copies of this single factor will suffice, while if an edge is present we shall need to have it available in $\mathbb{H}$ in both orientations. In any case the following amalgamation provides suitable factors.

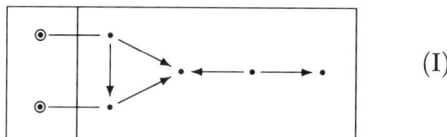
(I)

We still have to construct the two factors involved in this last diagram. They have a unique vertex in the first component. We shall call this vertex x, and display only the second component. All vertices realizing the type p over x will be labelled with a "p", and the remaining vertices will all realize the type q over x.

We consider first the factor:

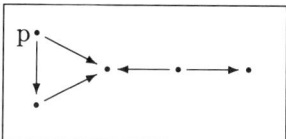

We obtain this from:

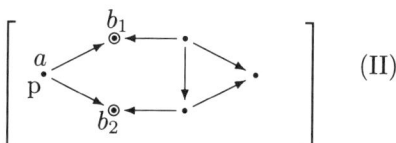
(II)

For the factor omitting b_2, fix the points xab_1 and look for the remaining three vertices in the associated 2-tournament $(x^q a^{\perp\prime} b, x^q (ab)^\perp)$ with at most two cross types. This reduces the problem to the construction of:
$$[\underset{p}{\bullet} \longrightarrow \bullet \longleftarrow \bullet \longrightarrow \bullet]$$

If we apply Lemma 6.5 to (x^p, x^q) then we need only construct the subconfigurations:

$$\boxed{p\bullet \longrightarrow \bullet \quad \bullet} \qquad \boxed{p\bullet \longrightarrow \bullet \longleftarrow \bullet}$$

The first is covered by Proposition 20 and the second by the dual of Case 1. Notice for future reference that the same argument yields the following configuration.
$$[\underset{p}{\bullet} \longrightarrow \bullet \longleftarrow \bullet \longleftarrow \bullet]$$

For the other factor in (II), omitting b_1, we work in (x^p, x^q). Lemma 6.6 was prepared for this purpose and reduces the problem to configurations already obtained.

We still have to construct the companion factor in (I):

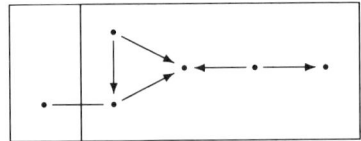

We use the following amalgamation, which either gives the necessary factor or delivers the 1-type $\mathbb{A}$ itself.

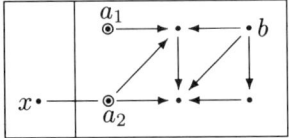

As $x^q \simeq \Gamma_2$, the factor omitting a_2 is assured. For the factor omitting a_1, we work in the 2-directed graph $\mathbb{H}^* = (x^p b^\perp, x^q b')$. The desired configuration is then:

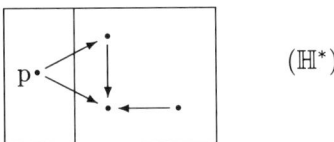
$(\mathbb{H}^*)$

As $x^q \simeq \Gamma_2$, we find also that $H_2^* \simeq \Gamma_2$. Thus the dual of Lemma 6.5 applies with edges representing $\perp$ and nonedges representing domination from H_1 to H_2. This reduces the problem to two configurations:

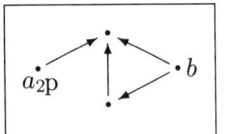 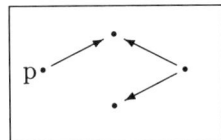

We have treated the second configuration already, and for the first, if we work over xa_2 this reduces to: $\boxed{p\bullet \longrightarrow \bullet \longleftarrow \bullet \longrightarrow \bullet b}$

which has occurred above. □

6.8. Lemma 6.4

We shall now prove Lemma 6.4. We continue to rely on Lemmas 6.5, 6.6 to provide a reduction mechanism for troublesome configurations.

PROOF. $\mathbb{H}$ is our homogeneous 2-directed graph satisfying the hypotheses of MT 2.2, and $\mathbb{A} = (\{x\}, 2 \cdot L(2))$ is a 1-type to be embedded in H. We shall let the two copies of $L(2)$ in A_2 be a_1, b_1 and a_2, b_2, and we shall let the type of x over a_1, a_2, b_1, b_2 be denoted p_1, p_2, q_1, q_2 respectively. Recall that there are at most three distinct cross types available.

Case 1. $p_1 = q_1$. Work over a_2, b_2, and prove nondiagonality by variation of parameters. Proposition 20 provides an adequate supply of cross types.

Case 2. p_1, p_2, q_1 all distinct.

Here a cross amalgamation succeeds:

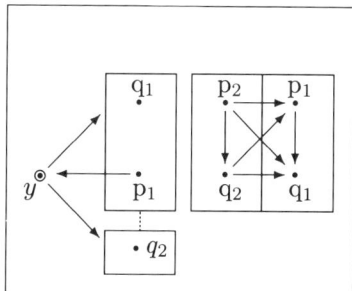

We only need to manufacture the factor omitting y, for which we amalgamate the two configurations shown:

(1) 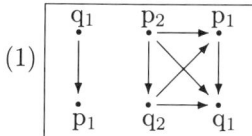 (2)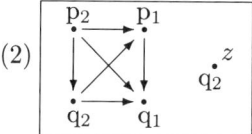

Now the first of these is the unique solution omitting $\mathbb{A}$ (and I_3) to an amalgamation diagram determing the type of the lefthand edge between vertices of type p_1, q_1. The factors of this amalgamation are both similar to the second of our configurations, so we confine our discussion to this one last configuration, where we have labelled a vertex z. In $(^{q_2}z, z^\perp)$ we can apply MT 2.1. This reduces the problem to configurations afforded by Proposition 20.

Case 3. $p_1 = q_2 = p$ and $p_2 = q_1 = q$.

We form:

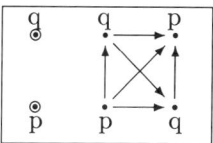

with factors of the type considered at the end of the last case.

Case 4. $p_1 = p_2 = p$, $q_1 = q_2 = q$.

We shall represent the type p by a solid edge, and we represent q by the absence of an edge. Our method will be to form suitable forms of the factors of:

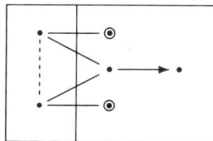

– either the single variant with no edge on the left, or the two variants with an oriented edge present. This will be a relatively long argument.

The amalgamation which furnishes suitable factors for this construction is:

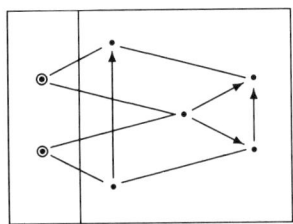

So we have to construct the following two configurations:

(3) 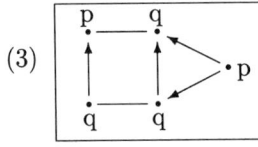   (4)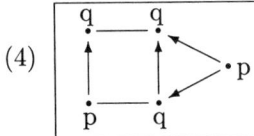

Configuration (4):

For (4) we use:

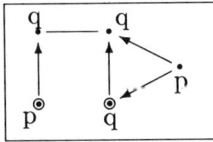

which results in an edge being inserted, or else produces $\mathbb{A}$ directly. The factors used in the last diagram are:

(5) 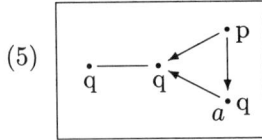   (6)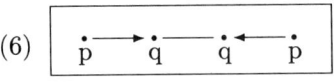

For (5) we apply Lemma 6.5 to (x^p, x^q), which reduces the problem to a 1-type over $I_1 + L(2)$, and another configuration to which Lemma 6.5 applies, and the two forms of the configuration omitting a, both of which are found in (6), with which we shall now deal.

For (6) use:

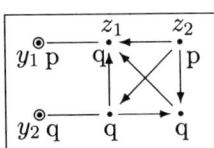

If no edge is inserted then $\mathbb{A}$ is produced.

Configuration (3):

At this point we have obtained (4), and we have also chosen the orientations of the edges in the process. As we proceed to construct the companion factor (3), orientations of all edges are to be taken as known in (3). We use the amalgamation:

6.8. LEMMA 6.4

For the factor omitting y_2, working over xy_1z_1, we have a 2-tournament $\mathbb{H}^* = (x^p y_1^{\perp \prime} z_1, x^q y_1^{\perp \prime} z_1)$ with at most two cross types, so the desired configuration embeds, assuming that the relevant cross type occurs and in addition the second component is nondegenerate (contains at least two vertices). In $\mathbb{H}$ we are asking for the following configurations, exhibited over x:

(7) 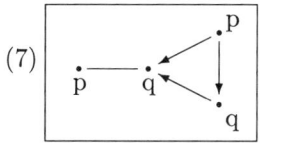 (8)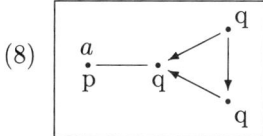

We shall discuss these below.

For the factor omitting y_1 we work over xy_2z_2 in the 2-directed graph $\mathbb{H}^* = (x^q y_2^{\pm} z_2', x^q y^{\perp} z_2')$. As the second component is a tournament, we are just asking that a certain cross type be realized, and that $\mathbb{H}^*$ be nondiagonal. Suppose on the contrary this structure is diagonal, and vary z_2. This gives rise to bijections defined in the 2-tournament $(x^p y_2^{\perp}, x^q y_2^{\perp})$ by two elements of the first component, which connect types over these elements in the second component. This can only happen in a nontrivial way if the latter 2-tournament is diagonal; but it has only two cross types. So all these bijections are trivial, which means that the bijection defined using xy_2z_2 could be defined just from xy_2. So to conclude it suffices to find the following two configurations, which represent the existence of a suitable cross type, and nondiagonality:

(9) 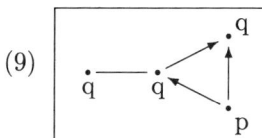 (10)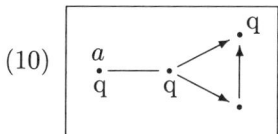

Thus we have seen at this stage that it suffices to deal with (7-10). Now (8) and (10) can be reduced by variation of parameters to configurations known to embed in $\mathbb{H}$. We still have to obtain the configurations (7, 9).

For (9) we work over x and apply Lemma 6.5. For (7) we form:

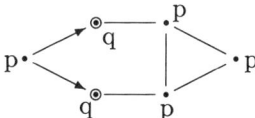

with factors:

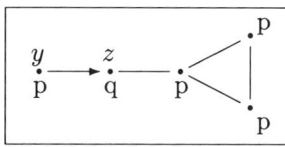

Working in the 2-tournament $(x^p y^{\perp} z^{\pm}, x^p y^{\perp} z^{\perp})$, which has at most two cross types, we see that it suffices to check that both cross types occur. In terms of $\mathbb{H}$ this means that we require all forms of:

(11) 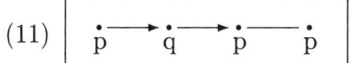 (12)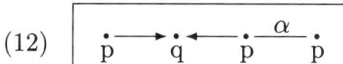

We work in $\mathbb{H}^* = (x^q, x^p)$. We need to show that $\mathbb{H}^*$ realizes certain 1-types over $I_1 + L(2)$. This is an analog of Proposition 20.

Configuration (11) corresponds to Case 1 in the proof of Proposition 20, which was handled by a cross amalgamation there. The configurations (1,2) of that argument correspond to the following in $\mathbb{H}$:

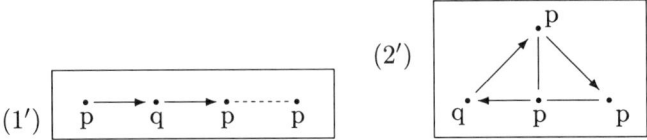

We make a form of (1′) from the factors:

Since (14) represents a 1-type over $I_1 + L(2)$, it embeds in $\mathbb{H}$ by Proposition 20. We shall consider (13) further below.

For (2′) we use:

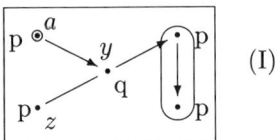

where the orientation of yz is selected in the construction of the factor omitting a, from subfactors:

(15) [diagram] (16) [diagram]

Here (16) is known to be available, while (15) requires further discussion.

The other factor in (I) is either (13) or:

(17) [diagram: p• → • ← •p with q]

which is covered by Lemma 6.3.

So the construction of (11) reduces to (13) and (15). We now consider (12), which corresponds to Case 3 or its dual in the proof of Proposition 20. Tracing through the argument given there, we come to the configuration:

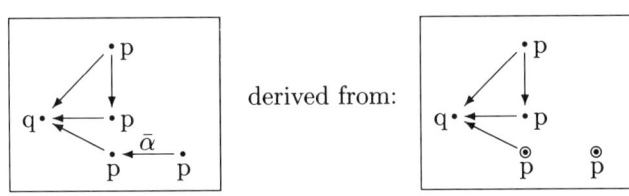

derived from:

with factors:

6.8. LEMMA 6.4

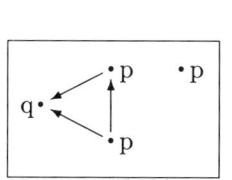 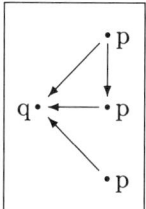

derived by variation of parameters.

We still have (13) and (15) to deal with:

Suppose that $\mathbb{H}$ omits (13). We use the following amalgamation:

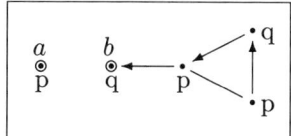

which produces $\mathbb{A}$ or (13). Either form of the factor omitting b will be obtained by applying MT 1.2 to $({}^p a, a^\perp)$. The other factor is afforded by:

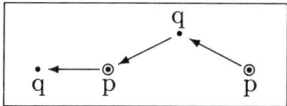

Here one factor is afforded by Lemma 6.3 and the other factor is:

(18)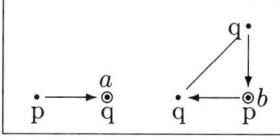

We get (18) (or (11), or $\mathbb{A}$ directly) using:

Some form of the factor omitting a is obtained from:

and the factor omitting b is obtained by variation of parameters.

Finally we show that $\mathbb{H}$ embeds:

(15)

We use:

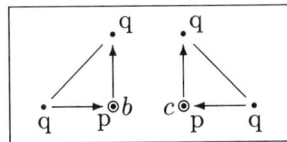

As this is quite symmetrical it suffices to get one factor, and as the orientation of the edge linking two vertices of type q still remains to be determined, this is easy. □

6.9. Lemma 6.5

In the proofs in this section, all our amalgamations involve 2-directed graphs which only involve a certain fixed pair of cross types, p, q. Accordingly we observe the convention that the cross type p is indicated by a solid edge, and the cross type q is indicated by the absence of an edge.

We first treat two special cases in which the configuration

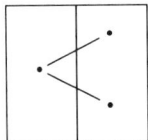

is omitted.

LEMMA 6.7. *Let $\mathbb{H}$ be a homogeneous 2-directed graph with $H_2 \simeq \Gamma_2$, omitting the configurations:*

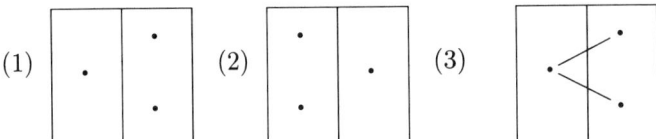

and embedding:

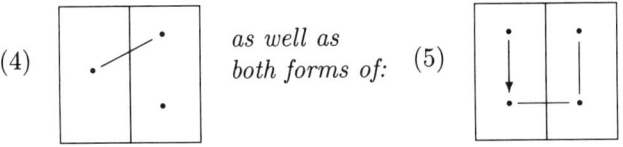

Then $\mathbb{H}$ embeds:

PROOF. We shall use:

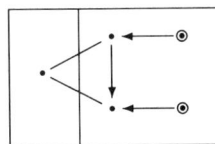

with the factors:

6.9. LEMMA 6.5

(8) (9)

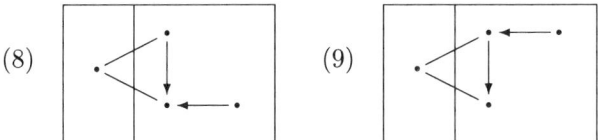

Assuming that $\mathbb{H}$ omits either (6) or (7), we shall force (8) into $\mathbb{H}$, using:

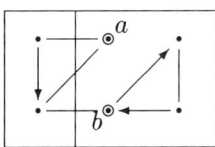

Here some form of the factor omitting a may be obtained from the two forms of (5). The other factor is:

(10)

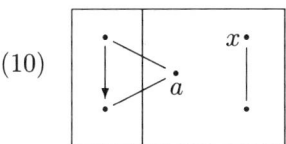

Over ax, we claim that a certain 2-tournament is nondiagonal. In the contrary case, varying a we find $(a_1 a_2)$-definable bijections in $^q x$. But $(^q x, x^\perp)$ is itself a 2-tournament, and by the classification of homogeneous 2-tournaments, these bijections are trivial. Thus if (10) is omitted, so is:

(11)

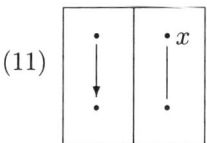

But $^q x$ is a tournament while $'x, x' \simeq \Gamma_2$. Thus $\mathbb{H}$ embeds (11), and hence also (8). Since $\mathbb{H}$ embeds (8) we may now suppose:

$$\mathbb{H} \text{ omits (9).}$$

We now show that $\mathbb{H}$ embeds (7), using the diagram:

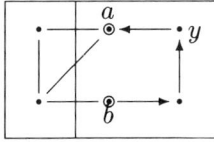

with the factors:

(12) (13)

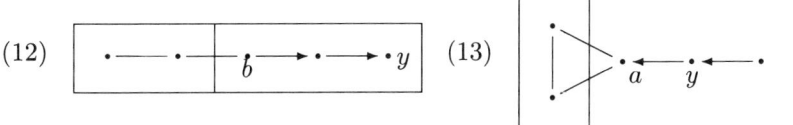

If (7) is omitted, (12) is a consequence of:

For (13) we work over ay. We claim that $(^pa\,^qy, a^\perp\,'y)$ is nondiagonal. Otherwise vary ay and study the resulting bijections inside H_2. This reduces the problem to:

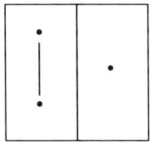

which occurs in (5). Thus $\mathbb{H}$ embeds (7).

Now we turn to (6). We use:

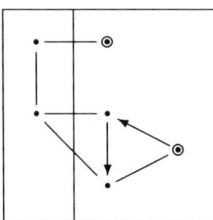

with the factors:

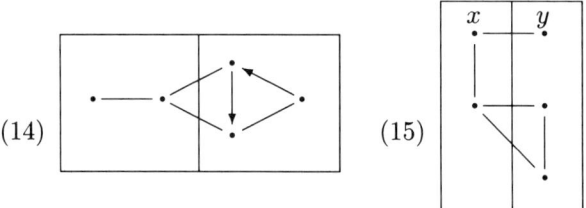

A form of (14) is readily obtained. We consider (15). Over xy we must show the corresponding 2-tournament is nondiagonal. If it is diagonal, vary x. We then find (x_1, x_2)-definable bijections inside $(^py, y^\perp)$. By the classification of homogeneous 2-tournaments, these are trivial. So $(^qy, y^\perp)$ is itself diagonal, that is $\mathbb{H}$ omits:

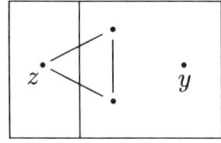

Vary the parameters to conclude. □

LEMMA 6.8. *Let $\mathbb{H}$ be a homogeneous 2-directed graph with $H_2 \simeq \Gamma_2$ which omits:*

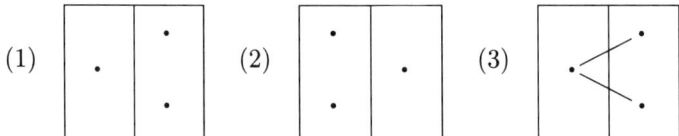

and embeds:

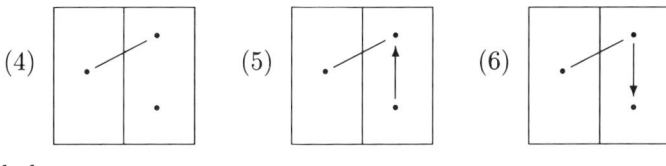

Then $\mathbb{H}$ *embeds:*

PROOF. By the preceding Lemma we may suppose that $\mathbb{H}$ omits a form of each of the following configurations:

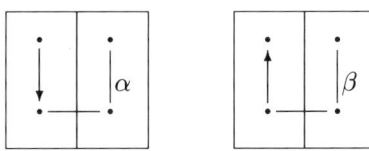

Here $\beta = \bar{\alpha}$ is the orientation opposed to α. Since we can reverse the orientations of the edges in H_1 without affecting anything, we take these configurations to be as shown:

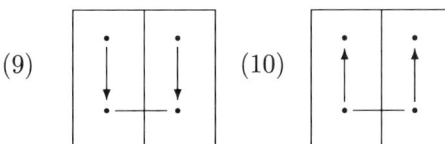

and then $\mathbb{H}$ embeds the alternate forms shown:

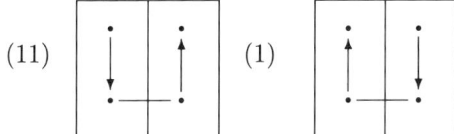

We consider the structure of $\mathbb{H}^* = ({}^q y, y')$. H_1^* is a tournament and $H_2^* \simeq \Gamma_2$. Furthermore $\mathbb{H}^*$ embeds:

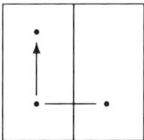

In particular H_1^* is infinite. If $H_1^* \simeq \mathbb{Q}$ then the set of realizations of any 1-type in H_1^* over a point in H_2^* is dense in H_1^*. This yields (9). If $H_1^* \simeq S(2)$ or T^∞ then ${}^q y$ contains $L(3)$ and C_3 and hence an amalgamation of the form:

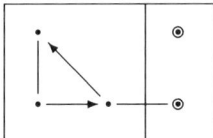

will produce (9) or (10), a contradiction. $\square$

LEMMA 6.9. *Let* $\mathbb{H}$ *be a homogeneous 2-directed graph with* $H_2 \simeq \Gamma_2$ *which omits:*

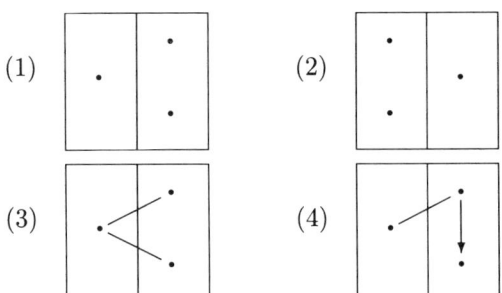

and embeds:

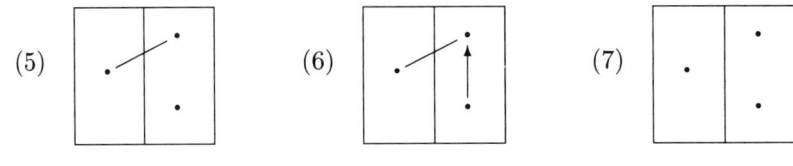

Then ℍ embeds:

PROOF. For $x \in H_1$, our claim is that (x^p, x^q) is not shuffled of type $(\mathbb{Q}, \mathbb{Q})$. It suffices to show that ℍ embeds:

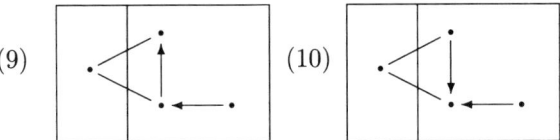

For (9) we use the amalgamation:

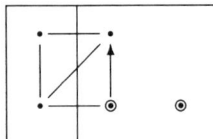

with factors

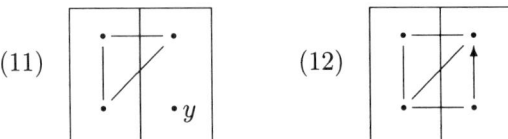

We have (11) since $(^q y, y^\perp)$ has at most two cross types and its second component is a tournament. Note also that $|^q y| > 1$ for similar resons. We force (12) by

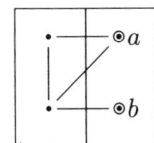

Here the factor omitting b is contained in (11). We get some form of the factor omitting a from

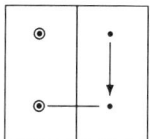

Now we consider (10). We note first that if $\mathbb{H}$ embeds the following:

(13) 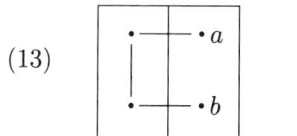 (14)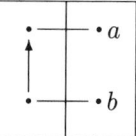

then the following amalgamation forces (10):

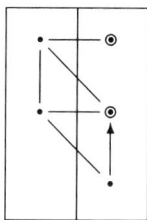

Now (13, 14) are the unique solutions to amalgamation problems determining the type of (ab). These have as their factors:

(15) 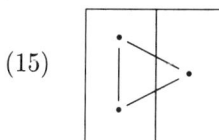 (16)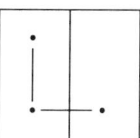

Configuration (15) is not a problem and for both forms of (16) we use

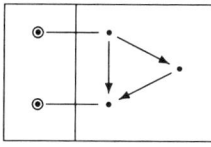

whose factors we clearly have. □

Now we shall prove Lemma 6.5. So we fix a homogeneous 2-directed graph $\mathbb{H}$ with $H_2 \simeq \Gamma_2$ which omits:

(1) 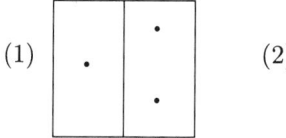 (2)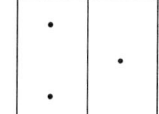

and embeds:

(3) 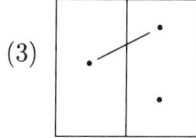 (4)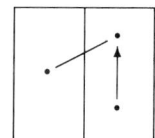

We must show that H embeds:

(5)  (6)

PROOF. By the preceding two lemmas we may suppose that $\mathbb{H}$ embeds:

(7)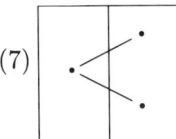

If H embeds:

(8)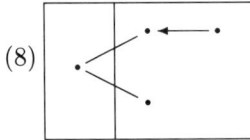

then use:

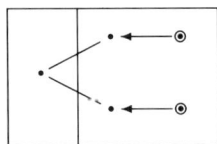

to conclude. Assume therefore that $\mathbb{H}$ omits (8).
 Observe that (8) is a consequence of:

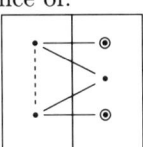

and that suitable factors for this diagram are produced by:

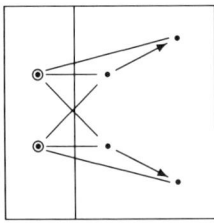

The factors here are both isomorphic with:

(9)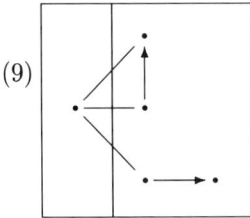

As (8) is omitted, (9) is afforded by:

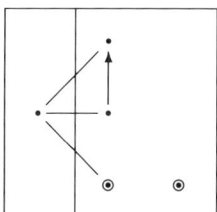

with the factors:

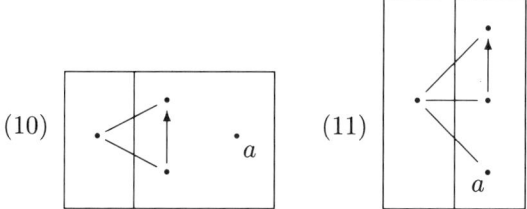

If one of the corresponding 2-directed graphs over a is diagonal, then varying the parameter a produces a contradiction. □

6.10. Lemma 6.6

The lemma states that if $\mathbb{H}$ is a homogeneous 2-directed graph in which H_1 omits I_∞, $H_2 \simeq \Gamma_2$, and $\mathbb{H}$ omits the configurations:

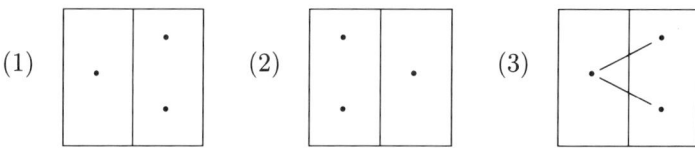

and embeds:

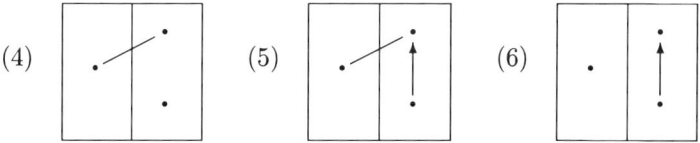

Then $\mathbb{H}$ also embeds:

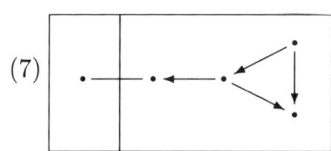

PROOF. If $\mathbb{H}$ omits (7) then for $x \in H_1$, (x^p, x^q) is a 2-tournament which omits a certain 1-type over $L(3)$ while realizing all its restrictions over copies of $L(2)$, by Lemma 6.5 and variation of parameters. Hence (x^p, x^q) is shuffled of type $(S(2), S(2))$.

For $a \in H_2$, $(^q a, a^\perp)$ is a 2-tournament whose second component is isomorphic with T^∞. For $x \in {}^q a$, $x^p \cap a^\perp$ is a subtournament of $S(2)$. Hence $(^q a, a^\perp)$ must either be diagonal with respect to the type p, or omit the cross type p, by the classification of homogeneous 2-tournaments. So $|x^p \cap a^\perp| \leq 1$. Similarly $x^q \cap a^\perp| \leq 1$, and hence $|x^q \cap a^\perp| = 1$, and $|x^p \cap a^\perp| = 0$, that is: $\mathbb{H}$ omits (4), a contradiction. □

With this, the classification of the homogeneous directed graphs omitting I_∞ is complete.

CHAPTER 7

Homogeneous directed graphs embedding I_∞

7.1. The classification theorem

To complete the classification of the homogeneous directed graphs, we must still deal with primitive homogeneous directed graphs embedding I_∞. There are two exceptional directed graphs in this category: $\mathcal{P}$ and the related directed graph $\mathcal{P}(3)$. In addition there is one general construction. If $\mathcal{T}$ is a set of finite tournaments closed under the formation of subtournaments, and A a finite directed graph, we shall say that A is $\mathcal{T}$-constrained if every tournament embedding in A is isomorphic with some tournament in the set $\mathcal{T}$. Let $\mathcal{A}(\mathcal{T})$ be the set of all finite $\mathcal{T}$-constrained directed graphs A. Then $\mathcal{A}(\mathcal{T})$ is an amalgamation class, and the associated homogeneous directed graphs will be denoted $\Gamma(\mathcal{T})$, and will be said to be $\mathcal{T}$-generic. We recall that there are $2^{\aleph_0}$ countable homogeneous directed graphs of this type.

THEOREM 7.1. *Let H be a (countable) primitive homogeneous directed graph embedding I_∞, and let $\mathcal{T}$ be the class of finite tournaments embedding in $\mathbb{H}$. Then unless H is $\mathcal{P}$ or $\mathcal{P}(3)$, H is $\mathcal{T}$-generic.*

The proof is quite similar to the one given in Chapter 4. The main theorem will consist of three parts, to be proved by simultaneous induction, with one part being the theorem that concerns us, while the other parts are included to keep the inductive argument moving.

NOTATION 4. *If $\mathcal{T}$ is a set of finite tournaments (usually a finite set for our current purposes) then $\mathcal{A}(\mathcal{T})$ will denote the set $\{I_n : n < \infty\} \cup \{I_1 + P_3\} \cup \mathcal{T}$, where P_3 is the oriented path of length two (and order 3).*

THEOREM 7.2 (MT 1). *Let $\mathcal{T}$ be a set of finite tournaments, closed under the formation of subtournaments.*
Then:

(MT 1.$\mathcal{T}$) *Every $\mathcal{T}$-constrained graph A belongs to every amalgamation class of finite directed graphs containing $\mathcal{A}(\mathcal{T})$.*

Before stating the rest of the Main Theorem, we wish to point out that MT 1.$\mathcal{T}$, for $\mathcal{T}$ any downward-closed class of finite tournaments – or equivalently, for $\mathcal{T}$ ranging over all finite sets of finite tournaments closed downward – is essentially identical to the Classification Theorem for primitive homogeneous directed graphs embedding I_∞, with the important exception of the analysis of sporadic cases. The sporadic cases are dealt with in Proposition 24 below: a primitive homogeneous directed graph embedding I_∞ is either I_∞, $\mathcal{P}$, or $\mathcal{P}(3)$, or else embeds $I_1 + P_3$. Of course, if H does embed $I_1 + P_3$, and if $\mathcal{T}$ is the set of all finite tournaments embedding in H, then H embeds all elements of $\mathcal{A}(\mathcal{T})$ and hence we can apply Theorem 1.$\mathcal{T}$ to H, so that H embeds all $\mathcal{T}$-constrained finite directed graphs, and is $\mathcal{T}$-generic. The exceptional cases in Proposition 24 include I_∞, which from the point of view of our catalog was not considered exceptional since it is $\{I_1\}$-generic.

The other two parts of the main theorem concern homogeneous 2-directed graphs. Though the definition used in Chapters 5, 6 will serve as well here, in fact the notion can be understood more broadly in the remainder of the article as referring to structures with two components, such that each component is a directed graph, and with a finite but arbitrary set of binary relations available as cross types between the components. The important homogeneous 2-directed graphs will be those derived from ordinary directed graphs by selecting two 1-types over some finite set, and these of course have at most three cross types. The following terminology is very useful.

NOTATION 5. ($\mathbb{H}$-constrained 1-types.)

1. When speaking of 2-directed graphs, a 1-type over the directed graph A will mean a 2-directed graph $\mathbb{A}$ for which A_1 is a singleton and $A_2 \simeq A$ (or $A_2 = A$ if the context requires it). Typically there will be a specific set of cross types allowed, generally consisting of all those present in some 2-directed graph currently under consideration.

2. A 2-directed graph $\mathbb{H}$ will be said to be ample if it embeds all 1-types over I_n for all n (with the stated convention on cross types), and if, in addition, $I_1 + P_3$ embeds in H_2.

3. If $\mathbb{A}$ is a 1-type over A and $B \subseteq A$, then the restriction of $\mathbb{A}$ to B is the 1-type (A_1, B) over B induced by $\mathbb{A}$.

4. If $\mathbb{H}$ is a 2-directed graph and $\mathbb{A}$ is a 1-type over A, $\mathbb{A}$ will be said to be $\mathbb{H}$-constrained if for every tournament T contained in A, the restriction of $\mathbb{A}$ to T embeds in $\mathbb{H}$.

THEOREM 7.3 (MT 2). *Let $\mathcal{T}$ be a set of finite tournaments, closed under the formation of subtournaments. Then:*

(MT 2.$\mathcal{T}$) *If $\mathbb{H}$ is an ample homogeneous 2-directed graph and $\mathbb{A}$ is an $\mathbb{H}$-constrained 1-type over $\oplus_i B_i$, where each $B_i \in \mathcal{A}(\mathcal{T})$, then $\mathbb{A}$ embeds in $\mathbb{H}$.*

THEOREM 7.4 (MT 3). *Let $\mathcal{T}$ be a set of finite tournaments, closed under the formation of subtournaments. Then:*

(MT 3.$\mathcal{T}$) *If $\mathbb{H}$ is an ample homogeneous 2-directed graph and $\mathbb{A}$ is an $\mathbb{H}$-constrained 1-type over a $\mathcal{T}$-constrained directed graph, then $\mathbb{A}$ embeds in $\mathbb{H}$.*

We have formulated our theorems in terms of a finite set $\mathcal{T}$ of finite tournaments so that they can be proved by a suitable induction on $\mathcal{T}$. We may now adopt the convention, to remain in force until the very end, that $\mathcal{T}$ always denotes a finite set of finite tournaments. We associate two invariants with $\mathcal{T}$:

$$\mathrm{rk}\mathcal{T} = \max\{|T| : T \in \mathcal{T}\};$$

$$\deg \mathcal{T} = |\{T \in \mathcal{T} : |T| = \mathrm{rk}\mathcal{T}\}|.$$

We write $\mathcal{T}_0 < \mathcal{T}_1$ if $(\mathrm{rk}\mathcal{T}_0, \deg \mathcal{T}_0) < (\mathrm{rk}\mathcal{T}_1, \deg \mathcal{T}_1)$ in lexicographic order. We work primarily by induction on this ordering.

7.2. The main ingredients

We shall now give a more detailed outline of the proof of our classification theorem. We formulate five specialized propositions and four theorems that constitute the main steps in the derivation of our three main theorems. Typically the most specialized results depend on very explicit amalgamation arguments and then the general results follow quite formally.

PROPOSITION 24. *Let H be a primitive homogeneous directed graph embedding I_∞. Then either H is isomorphic with one of I_∞, $\mathcal{P}$, or $\mathcal{P}(3)$, or else H embeds $I_1 + P_3$.*

We shall use the notation $\mathcal{A}$ ent B discussed in section §1.3. In this case we also say that B is a *consequence* of $\mathcal{A}$. We shall generally write $\mathcal{A} \implies \mathcal{B}$ for $\& \mathcal{A} \implies \& \mathcal{B}$, that is, $\mathcal{A} \implies B$ for all $B \in \mathcal{B}$. This is very different from the equally important concept $\mathcal{A} \implies \bigvee \mathcal{B}$.

We make a general observation that will be applied tacitly later on. Let $\mathcal{A}$ be a set of finite directed graphs, B a finite directed graph, and suppose $\mathcal{A} \implies B + A$ for all $A \in \mathcal{A}$. Then $\mathcal{A} \implies B + A$ for all A such that $\mathcal{A} \implies A$. One way to see this is to consider a homogeneous directed graph H embedding all elements of $\mathcal{A}$, to fix a copy of B in H, and to consider $B^\perp$. In a similar vein, if $\mathcal{A} \implies A_1 + A_2$ when $A_1, A_2 \in \mathcal{A}$, then also $\mathcal{A} \implies A_1 + A_2$ when A_1, A_2 are consequences of $\mathcal{A}$; the previous result may be applied twice.

Our next result concerns the following construction. If A is a finite directed graph let H^+, H^- be the directed graphs on the vertex set $H \times \{0, 1\}$ with edges defined by $(a, 0) \longrightarrow (b, 0)$ iff $a \longrightarrow b$ and:

(+) $(a, 0) \longrightarrow (a, 1)$ for all a;
  or
(−) $(a, 1) \longrightarrow (a, 0)$ for all a,

respectively.

LEMMA 7.5. *Let $\mathcal{A}$ be a set of finite directed graphs such that for $A \in \mathcal{A}$ we have $\mathcal{A} \implies A^+$ and $\mathcal{A} \implies A + I_1$. Then $\mathcal{A} \implies A^+$ whenever $\mathcal{A} \implies A$.*

PROOF. This depends on the fact that if $\mathcal{A} \implies A$ then there is a finite tree of amalgamation diagrams, with the branching at each node corresponding to the possible completions of the diagram at the given node, and with the factors in each diagram coming from $\mathcal{A}$ or from the assumed results of the diagrams along earlier nodes, so that along each maximal branch of the tree, A is contained in some configuration associated with that branch. Compare [61].

In addition we require the following observation. Let A_1, A_2 be extensions of A_0, and consider the possible amalgams of $A_1^+ \cup (A_2^+ - A_2)$, $A_2^+ \cup (A_1^+ - A_1)$ over $A_0 + (A_1^+ - A_1) + (A_2^+ - A_2)$. This has factors of the forms $I_n + A_i^+$, and any solution contains A^+ for some amalgam A of A_1, A_2 over A_0.

Let $\mathcal{A} + I_n = \{A + I_n : A \in \mathcal{A}\}$ and similarly $\mathcal{A}^+ = \{A^+ : A \in \mathcal{A}\}$. Then from a tree of amalgamation diagrams witnessing $\mathcal{A} \implies A$ we derive a tree witnessing $\mathcal{A}^+ + I_n \implies A^+$ for some n, and by hypothesis every directed graph in $\mathcal{A}^+ + I_n$ is a consequence of $\mathcal{A}$. □

We may now state the remaining four propositions and three theorems needed for our argument.

PROPOSITION 25. *If $\mathcal{A}(\emptyset) \implies A$ then $\mathcal{A}(\emptyset) \implies L(2) + A$.*

PROPOSITION 26. *If $\mathcal{A}(\emptyset) \implies A$ then $\mathcal{A}(\emptyset) \implies A^+$ and A^-.*

PROPOSITION 27. *If $\mathbb{H}$ is an ample homogeneous 2-directed graph and $\mathbb{A}$ is an $\mathbb{H}$-constrained 1-type over $2 \cdot L(2)$, then $\mathbb{A}$ embeds in $\mathbb{H}$.*

PROPOSITION 28. *If $\mathbb{H}$ is an ample homogeneous 2-directed graph and $\mathbb{A}$ is an $\mathbb{H}$-constrained 1-type over P_3, then $\mathbb{A}$ embeds in $\mathbb{H}$.*

THEOREM 7.6. *Suppose that $T \in \mathcal{T}$, A is a finite $\mathcal{T}$-constrained directed graph containing T, $b \in A$, and every tournament contained in $A^* = A - \{b\}$ is smaller than T. Then $\mathcal{A}(\mathcal{T}) \implies A$.*

THEOREM 7.7. *Let $A = \{a, b\} \cup T$ be $\mathcal{T}$-constrained with T a tournament and $a \perp T$. Then any homogeneous directed graph embedding the elements of $\mathcal{A}(\mathcal{T})$ must also contain A.*

Here $\{a, b\} \cup T$ denotes a directed graph A with vertex set $V(A) = \{a, b\} \cup V(T)$, and with the tournament T as an induced directed subgraph.

THEOREM 7.8. *Let $\mathbb{H}$ be an ample homogeneous 2-directed graph and let $\mathbb{A}$ be an $\mathbb{H}$-constrained 1-type over $T_1 + T_2$ with $T_1, T_2 \in \mathcal{T}$. Then $\mathbb{A}$ embeds in $\mathbb{H}$.*

THEOREM 7.9. *If A is the disjoint sum of graphs in $\mathcal{A}(T)$ and A^* is a $\mathcal{T}$-constrained extension of A by a single point, then every homogeneous directed graph embedding the elements of $\mathcal{A}(\mathcal{T})$ must also contain A^*.*

We shall also index these last four theorems by $\mathcal{T}$: Theorems 7.6.$\mathcal{T}$, 7.7.$\mathcal{T}$, 7.8.$\mathcal{T}$, and 7.9.$\mathcal{T}$.

7.3. Structure of the proof

Formulating the nature of our inductive proof precisely is one of main steps in the argument. Indeed, the main difficulty was to find statements to be incorporated into the Main Theorem so that the induction can proceed. Accordingly we now give a precise formulation of the steps of the proof, so that we can subsequently treat them independently.

1. Prove the five propositions, 24 - 28. We shall need Propositions 26 and 28. The role of Proposition 24 has already been explained. Proposition 25 is used in the proof of Proposition 26 and Proposition 27 is used in the proof of Proposition 28.
2. Assume Proposition 26, Theorem 7.7.$\mathcal{T}_0$ and MT 3.$\mathcal{T}_0$ for $\mathcal{T}_0 < \mathcal{T}$. Prove Theorems 7.6.$\mathcal{T}$ and 7.7.$\mathcal{T}$.
3. Assume MT 3.$\mathcal{T}_0$ for $\mathcal{T}_0 < \mathcal{T}$ and Theorems 7.6.$\mathcal{T}$ and 7.7.$\mathcal{T}$. Prove Theorem 7.8.$\mathcal{T}$.
4. Assume Propositions 26 and 28, and Theorem 7.8.$\mathcal{T}$. Prove MT 2.$\mathcal{T}$.
5. Assume Propositions 26 and 28, and assume Theorem 7.7.$\mathcal{T}$, MT 2.$\mathcal{T}$, and MT 3.$\mathcal{T}_0$ for $\mathcal{T}_0 < \mathcal{T}$. Derive Theorem 7.9.$\mathcal{T}$.
6. Assume Proposition 26 and Theorem 7.9.$\mathcal{T}$. Prove MT 1.$\mathcal{T}$.
7. Assume MT 1.$\mathcal{T}$ and MT 2.$\mathcal{T}$. Prove MT 3.$\mathcal{T}$.

7.4. Steps 4, 6, 7. The Main Theorem

We shall carry out these seven steps in the order: 6, 4, 7; 1; 2, 5, 3. Steps 6, 4, 7 put the induction together, proving MT 1–3. Steps 2, 3, 5 prove the auxiliary theorems; they are deferred to the next chapter. Step 3 is the most delicate.

This will then complete the classification of all countable homogeneous directed graphs, modulo known classification results we have cited: the finite and imprimitive cases, the classification of homogeneous tournaments, and the partially ordered case.

7.4. Steps 4, 6, 7. The Main Theorem

We shall now carry out steps 4, 6, and 7 of our outline. With some abuse of language we may say that we are giving the derivation of the Main Theorem from our five propositions and three more technical theorems, or even from Proposition 28 and Theorem 7.8. That is the gist of the matter, but due to the inductive nature of the proof it is only very roughly correct. Indeed, matters have been organized so that this part of the proof is largely a formal exercise.

Step 6. Assume Proposition 26 and Theorem 7.9.$\mathcal{T}$. Prove MT 1.$\mathcal{T}$.

By convention $\mathcal{T}$ is a finite set of finite tournaments which is closed under the formation of subtournaments.

Our hypothesis is:

> If A is the disjoint sum of graphs in $\mathcal{A}(T)$ and A^* is a $\mathcal{T}$-constrained extension of A by a single point, then every homogeneous directed graph embedding the elements of $\mathcal{A}(\mathcal{T})$ must also contain A^*.

The desired conclusion is:

> Let $\mathcal{T}$ be a set of finite tournaments, closed under the formation of subtournaments. Then:
>
> [MT 1.$\mathcal{T}$] Every $\mathcal{T}$-constrained graph A belongs to every amalgamation class of finite directed graphs containing $\mathcal{A}(\mathcal{T})$.

Here Ramsey arguments will play a role.

DEFINITION 7.10. Let r be a nontrivial 2-type. A *Ramsey structure* of type r is a structure R which can be linearly orderd so that any pair of elements a, b with $a < b$ has the type r. If r is asymmetric, then the structure is linearly ordered by the type r, while if r is symmetric, the ordering is irrelevant. A Ramsey directed graph of type $\longrightarrow$ or $\longleftarrow$ is a linear tournament, and a Ramsey directed graph of type $\perp$ is a copy of I_n for some n.

DEFINITION 7.11. Let $\mathcal{T}$ be a finite set of finite tournaments, let $\mathcal{A}$ be an amalgamation class containing $\mathcal{A}(\mathcal{T})$, and let r be a 2-type. We define $\mathcal{A}^r$ to be the set of all $A \in \mathcal{A}$ satisfying:

(r) if $R \cup A$ is any extension of A by a finite r-Ramsey directed graph R such that $\{x\} \cup A$ is $\mathcal{T}$-constrained for all $x \in R$, then $R \cup A \in \mathcal{A}$.

In the notation the dependence of $\mathcal{A}^r$ on $\mathcal{T}$ is not shown, but in practice one normally knows which class $\mathcal{T}$ is under consideration, and more specifically one normally assumes $\mathcal{A} \supseteq \mathcal{A}(\mathcal{T})$.

The following is immediate and useful.

LEMMA 7.12. *If $\mathcal{A}$ is an amalgamation class, r is a nontrivial 2-type, and $\mathcal{T}$ is a finite set of finite tournaments, then $\mathcal{A}^r$ is an amalgamation class.*

The proof of the next theorem uses Theorem 7.9 and Lachlan's Ramsey argument, and is immediate modulo those ingredients:

LEMMA 7.13. *If $\mathcal{A}$ is an amalgamation class containing $\mathcal{A}(\mathcal{T})$ then there is a 2-type r for which $\mathcal{A}^r$ contains $\mathcal{A}(\mathcal{T})$.*

PROOF. If we suppose the contrary, then for every nontrivial 2-type r we have a graph $A_r \in \mathcal{A}(\mathcal{T})$ and a finite r-Ramsey extension $B_r = R_r \cup A_r$ of A_r such that $\{x\} \cup A_r$ is $\mathcal{T}$-constrained for all $x \in R_r$, while B_r is not in $\mathcal{A}$. Let $A = \oplus_r A_r$ (the disjoint union of three such directed graphs). We apply Theorem 7.9 to this graph: any $\mathcal{T}$-constrained extension of $n \cdot A$ by a single point will lie in $\mathcal{A}$, and hence by Lachlan's Ramsey argument, for n large this forces $R_r \cup A_r$ to be in $\mathcal{A}$ for some r, a contradiction. □

We may now carry out the formal portion of the proof of MT 1.$\mathcal{T}$.

PROOF. We prove by induction on n:

(1.$\mathcal{T}$.n) If $\mathcal{A}$ is an amalgamation class containing $\mathcal{A}(\mathcal{T})$, and H is a $\mathcal{T}$-constrained directed graph with n vertices, then $H \in \mathcal{A}$.

For $n = 1$ this is clear. For $n > 1$, with $\mathcal{A}$ and H fixed choose $x \in H$ and let r be a 2-type so that $\mathcal{A}(\mathcal{T}) \subseteq \mathcal{A}^r$. Let $H^* = H - \{x\}$. By induction on n, $H^* \in \mathcal{A}^r$, and setting $R_r = \{x\}$ we find that $H = R_r \cup H^*$ belongs to $\mathcal{A}$, as claimed. □

Step 4. Assume Propositions 26 and 28 and Theorem 7.8.$\mathcal{T}$. Prove MT 2.$\mathcal{T}$.

We shall also make use of Proposition 25, but this may be thought of as a special case of Proposition 26 (consider A^{+++}). We fix an ample homogeneous 2-directed graph $\mathbb{H}$. We must first prove a special case of MT 2.$\mathcal{T}$.

LEMMA 7.14. *If $T \in \mathcal{T} \cup \{P_3\}$ and $\mathbb{A}$ is a 1-type over $T + I_n$ whose restriction to T embeds in $\mathbb{H}$, then $\mathbb{A}$ embeds in $\mathbb{H}$.*

PROOF. We proceed by induction on n. Let $A_1 = \{x\}$, fix $y \in I_n \subseteq A_2$, and let $p = \text{tp}(xy)$. Identify y with an element of H_2, and let $\mathbb{H}^* = (^p y, y^\perp)$. Then $\mathbb{H}^*$ is again homogeneous, and we claim that it is ample. $\mathbb{H}^*$ realizes all 1-types over all I_n, and H_2^* embeds $I_1 + P_3$ by Proposition 25. $\mathbb{H}^*$ realizes the same 1-types over each $A \in \mathcal{T}$ as $\mathbb{H}$ does, by an application of Theorem 7.8 with $T_2 = I_1$. By induction $\mathbb{A} - \{y\}$ embeds in $\mathbb{H}^*$ and hence $\mathbb{A}$ embeds in $\mathbb{H}$. □

LEMMA 7.15. *H_2 contains $P_3 + T$ for every $T \in \mathcal{T}$.*

PROOF. We may suppose $L(2) \in \mathcal{T}$. Fix $c \in H_2$ and let $\mathbb{H}^c$ be $(c', c^\perp)$. It suffices to embed $\mathbb{A} = (a, I_1 + T)$ in $\mathbb{H}^c$ with $a \longrightarrow I_1$, $a \perp T$. Now $\mathbb{H}^c$ is ample, using Propositions 25 and 26, so applying Theorem 7.8.$\mathcal{T}$ to $\mathbb{H}^c$, it suffices to check that $\mathbb{A}$ is $\mathbb{H}^c$-constrained. For this we need $L(2) + T$ in H_2, which follows from Theorem 7.8.$\mathcal{T}$. □

Now we treat the general case. We shall prove MT 2.$\mathcal{T}$ in the following form:

(2.$\mathcal{T}$.n) If $A = \oplus_i B_i$ is the direct sum of n graphs B_i, all lying in $\{P_3\} \cup \mathcal{T}$, then every $\mathbb{H}$-constrained 1-type $\mathbb{A}$ over A embeds in $\mathbb{H}$.

PROOF. Our proof proceeds by induction on n. For $n = 1$ the claim is vacuous when $A \in \mathcal{T}$, and for $A = P_3$ our claim is stated as Proposition 28, which we are presently assuming.

For $n > 1$ let p be the restriction of $\mathbb{A}$ to B_1 and let $\mathbb{H}^* = (^p B_1, B_1^\perp)$, taking $B_1 \subseteq H$. We claim that $\mathbb{H}^*$ is ample. This follows from Lemmas 7.14 and 7.15.

Now since $\mathbb{H}^*$ is ample, and has the same 1-types over elements of $\mathcal{T}$ as $\mathbb{H}$ does, an application of the inductive hypothesis to $\mathbb{H}^*$ embeds $\mathbb{A} - B_1$ into $\mathbb{H}^*$, and hence embeds $\mathbb{A}$ into $\mathbb{H}$. □

Comment (for use in Chapter 8) We need a sharper version of the statement in Chapter 8, during the proof of Lemma 8.5. Suppose we know Propositions 26 and 28, and $\mathbb{H}$ is an ample homogeneous 2-directed graph such that Theorem 7.8.$\mathcal{T}$ applies to every 2-directed graph of the form $\mathbb{H}^* = (^p A, A^\perp)$ with $A \leq H_2$ and with p a 1-type over A which is realized in H_1. We can then try to repeat the argument given above to prove that the statement of MT 2.$\mathcal{T}$ holds for $\mathbb{H}$. The only problem that arises lie in the proof of Lemma 7.15, where the 2-directed graph introduced is not of the form to which our hypothesis applies. We conclude:

> *Assume Propositions 26 and 28. Let $\mathbb{H}$ be an ample homogeneous 2-directed graph. By a* derived 2-directed graph *we mean a 2-directed graph of the form $\mathbb{H}^* = (^p A, A^\perp)$ with $A \leq H_2$ and with p a 1-type over A which is realized in H_1. Suppose that every 2-directed graph derived from $\mathbb{H}^*$ satisfies the conclusion of Theorem 7.8.$\mathcal{T}$ and also satisfies: $P_3 + T$ embeds in H_2^* for $T \in \mathcal{T}$. Then $\mathbb{H}$ satisfies the conclusion of MT2.$\mathcal{T}$.*

Step 7. Assuming MT 1.$\mathcal{T}$ and MT 2.$\mathcal{T}$, prove MT 3.$\mathcal{T}$.

PROOF. We fix an ample homogeneous 2-directed graph $\mathbb{H}$. We want to prove that every $\mathbb{H}$-constrained 1-type over a $\mathcal{T}$-constrained directed graph will embed in $\mathbb{H}$. As in the proof of Lemma 7.13, we may deduce from MT 2.$\mathcal{T}$ that there is a 2-type r realized in H_1 such that for every finite $\mathbb{H}$-constrained configuration (R, A) with R r-ramsey and $A = \oplus B_i$ constructed from $B_i \in \mathcal{A}(\mathcal{T})$, (R, A) embeds in $\mathbb{H}$. If we let $\mathcal{A}$ be the class of all finite directed graphs A which satisfy this extension property, then $\mathcal{A}$ is an amalgamation class containing $\mathcal{A}(\mathcal{T})$. By MT 1.$\mathcal{T}$ it follows that $\mathcal{A}$ contains all $\mathcal{T}$-constrained graphs, which yields our claim. □

7.5. Step 1. Proposition 24: P_3

The treatment of Step 1 occupies the next three sections. In this section we prove Proposition 24, which disposes of the sporadic cases.

LEMMA 7.16. *If the primitive homogeneous directed graph H embeds I_∞ and $L(2)$, and is not isomorphic to $\mathcal{P}$, then H embeds P_3.*

PROOF. If H is a partial ordering then use the classification obtained in [**82**]. We assume that H is not a partial ordering, but omits P_3. Then H embeds C_3. Furthermore, since H omits P_3 and is primitive, H embeds $[I_1, I_2]$ or its dual, and we may suppose the first alternative occurs. The remainder of the argument traces through the effects of the following amalgamations in H. The first two diagrams provide one factor of the fourth one, and the third diagram produces the other factor.

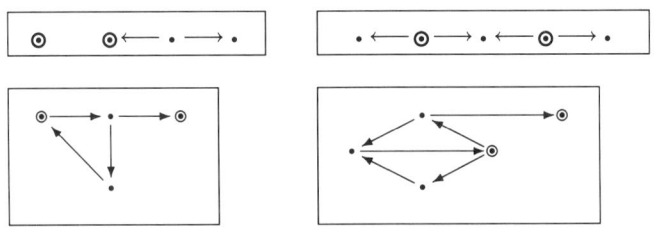

LEMMA 7.17. *Let H be a primitive homogeneous directed graph embedding I_∞ and P_3. Then H embeds $I_2 + L(2)$.*

PROOF. Suppose that H omits $I_2 + L(2)$, and fix $a \in H$. On $a^\perp$, $\perp$ is an equivalence relation with infinite classes. Let C be one such class. For $x \in a'$, we have $|x^\perp \cap C| \le 1$. There is a type $p \in \{\longrightarrow, \longleftarrow\}$ such that for $x \in a'$, $x^p \cap C$ is infinite whenever $x^\perp$ meets C. We claim:

(1)   For $x, y \in a'$, if $x^\perp$ and $y^\perp$ meet C, then $x^p \cap y^p \cap C$ is infinite.

Suppose (1) fails for some pair $x, y \in a'$. Let q be the other asymmetric 2-type. Then $(x^p \cap y^q) \cap C$ is infinite. Hence $x^\perp$, x', $'x$ all meet C, with $x' \cap C$ and $'x \cap C$ infinite. It follows that the same holds for any choice of $x \in a'$ and $C \in a^\perp/\perp$.

Fix $b \in C$. Define $f : a' \cap b' \to C$ by: $x \perp f(x)$. For $x, y \in a' \cap b'$, $\mathrm{tp}(xy)$ determines the types of $f(x)y$ and $xf(y)$, which may however be chosen arbitrarily in $\{\longrightarrow, \longleftarrow\}$, giving four nontrivial 2-types in $a' \cap b'$, a contradiction. In more detail, if $b, b_1, b_2 \in C$ are distinct and $r \in \{\longrightarrow, \longleftarrow\}$, then we can find $x \in a' \cap b'$ with $f(x) = b_1$ and with $\mathrm{tp}(xb_2) = r$.

Thus (1) holds. Now we claim:

(2)   For $x \in a'$ we have $|x^\perp \cap a^\perp| > 1$.

If we suppose the contrary then there is a surjection $f : a' \to a^\perp$ defined by $f(x) \perp x$, and there are then at least four nontrivial 2-types realized in $a' \times a^\perp$: $f(x) = y$; $f(x) \perp y$; $f(x) \longrightarrow y$; $f(x) \longleftarrow y$. Thus (2) must hold.

Now we reach a contradiction. Choose $x \in a'$, $b_1, b_2 \in x^\perp \cap a^\perp$ with $b_1 \longrightarrow b_2$. Choose $b_3 \perp b_2$ in $a^\perp$, and choose $y \in a' \cap b_1^\perp \cap b_2^\perp$, by homogeneity. Choose $c_1, c_2 \in a^\perp \cap x^p \cap y^p$ with $c_1 \perp b_1$, $c_2 \perp b_2$. Then $\mathrm{tp}(xyc_1/a) = \mathrm{tp}(xyc_2/a)$, but $x^\perp \cap y^\perp \cap a^\perp$ meets $c_1^\perp$ and not $c_2^\perp$, a contradiction. □

We now prove Proposition 24, which states that a primitive homogeneous directed graph H which is not one of I_∞, $\mathcal{P}$, or $\mathcal{P}(3)$ must embed $I_1 + P_3$.

PROOF. By Lemma 7.16 we may suppose that H embeds P_3 and omits $I_1 + P_3$, and we must prove that H is then $\mathcal{P}(3)$. By the preceding lemma H embeds $I_2 + L(2)$. So for $x \in H$, either $x^\perp \simeq \infty \cdot T$ with T homogeneous, or $x^\perp \simeq \mathcal{P}$.

If $x^\perp \simeq \infty \cdot T$, then for $y \in x^\perp$, $y^\perp$ will consist of all but one component of $x^\perp$, and a component containing x. As H is primitive, the equivalence relation defined by: "$x^\perp, y^\perp$ coincide up to a finite number of components" must hold for any $x, y \in H$. In particular if A is a copy of P_3 in H, then $A^\perp$ is nonempty, a contradiction.

Our conclusion is that $x^\perp \simeq \mathcal{P}$. Thus H also omits $I_1 + C_3$. We now argue that none of the following amalgamation diagrams can be formed in H:

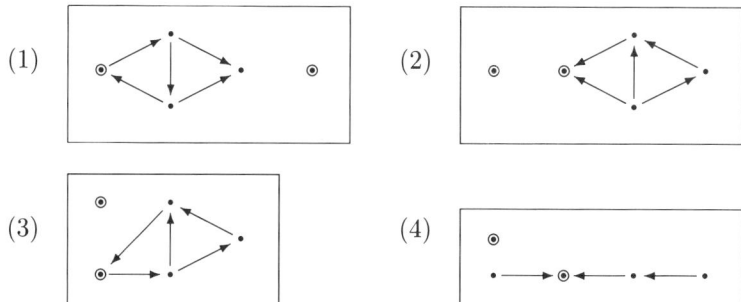

The first three yield either a contradiction, the fourth diagram, or the dual of the fourth diagram. The fourth diagram yields the following, which in turn produces a contradiction:

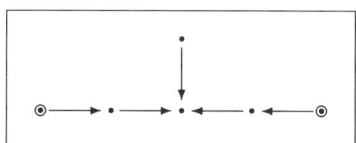

Now in each of (1-4), as well as their duals, there is only one factor which can be omitted by H. As the factor occurring in (3) is self-dual, this yields seven configurations that H is forced to omit, in addition to $I_1 + P_3$ and $I_1 + C_3$. One of the forbidden configurations is $[I_1, P_3]$, so for $x \in H$ x' must embed $\mathcal{P}$ (as $x^\perp \simeq \mathcal{P}$) and omit P_3, forcing $x' \simeq \mathcal{P}$ by Lemma 7.16. Similarly $'x \simeq \mathcal{P}$. Consequently H omits $[I_1, C_3]$ and $[C_3, I_1]$. We have now identified 11 forbidden configurations, all of order 4, and we claim that the structure of H is now determined explicitly.

Let $\mathbb{H}$ be the three-directed graph $(x^\perp, x', 'x)$. Our claim is that this is isomorphic with the 3-directed graph we used to construct $\mathcal{P}(3)$. The eleven factors omitted by H impose 38 restrictions on $\mathbb{H}$, each restriction corresponding to a forbidden configuration on four vertices, now thought of as a configuration of order three over one of the four vertices. Eight of the configurations involved produce four distinct constraints in this manner, while three of them ($I_1 + C_3$, $[I_1, C_3]$, $[C_3, I_1]$) produce only two constraints apiece. We may check directly that after transforming $\mathbb{H}$ to a 3-directed graph $\mathbb{P}^*$ by shifting the types in $b^i \times b^j$ by $i - j$, these constraints will signify that $\mathbb{P}^*$ is a 3-partitioned partial order. However it is simpler to note that $\mathcal{P}(3)$ satisfies all of the conditions imposed on H and hence its associated 3-directed graph $\mathbb{P}^*$ satisfies our 38 constraints of order 3. On the other hand there are exactly 38 excluded configurations of order three in $\mathbb{P}^*$, corresponding to 27 variants of P_3 and 11 variants of C_3, which is the number of distinct ways of distributing the three vertices involved among the three 1-types, taking into account the automorphisms of C_3. Thus the 38 conditions we have found must be the ones that force $\mathbb{P}^*$ to come from a 3-partitioned partial order. It is clear that each component of $\mathbb{H}^*$ is dense. Let H^* denote the partial order formed from $\mathbb{H}^*$ by shifting cross types.

We still need to check that each component of $\mathbb{H}^*$ is dense in $\mathcal{P}$. As $\mathbb{P}^*$ is homogeneous, it suffices to show that $\mathbb{P}^*$ embeds the configurations $(c_1, c_2, c_3) \simeq L(3)$ with $c_1, c_3 \in P_i$ and $c_2 \in P_{i+1}$, for each $i \in \mathbb{Z}/3\mathbb{Z}$. That is, we must check that three specific configurations on four vertices embed into H. Taking into account the known restrictions on H, we see that the following diagrams contain the necessary amalgamations:

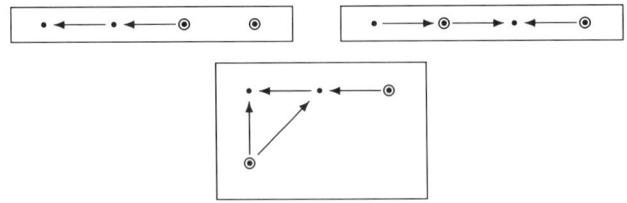

7.6. Step 1, Proposition 25: adding $L(2)$

We shall prove Proposition 25 in this section: if $\mathcal{A}(\emptyset) \implies A$, then $\mathcal{A}(\emptyset) \implies A + L(2)$.

LEMMA 7.18. *If $\mathcal{A}(\emptyset) \implies A$ then $\mathcal{A}(\emptyset) \implies I_1 + A$.*

PROOF. All we need to show is that $I_1 + P_3 \implies I_2 + P_3$. If H is a homogeneous directed graph embedding $I_1 + P_3$, but omitting $I_2 + P_3$, then for $a \in H$, $a^\perp$ is primitive, using Lemma 7.17, and isomorphic with $\mathcal{P}(3)$ by Proposition 24. Now the following amalgamation diagram will be sufficient, as it results in $I_2 + P_3$ no matter how it is completed:

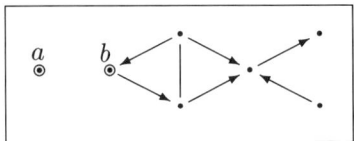

As H contains $I_1 + \mathcal{P}(3)$, the factor omitting b must occur in H regardless of the orientation still to be determined along one edge. The other factor may then be manufactured by an amalgamation determining this orientation. The factors needed for this are:

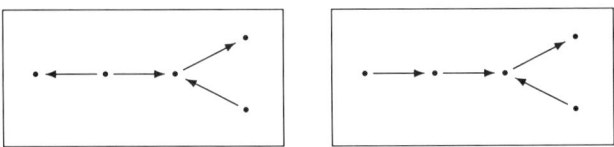

Because $x^\perp \simeq \mathcal{P}(3)$ for $x \in H$, the two diagrams following have unique solutions, containing the desired factors.

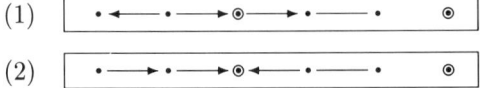

For the same reason (3,4) below also result in the insertion of an edge in each case, giving the missing factor in (1,2) respectively. It remains to be seen that (3, 4) can be set up in H.

If e.g. (4) cannot be constructed, then H omits the first diagram following, and the succeeding two diagrams may be used to get $I_2 + P_3$.

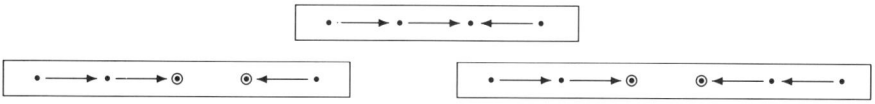

$\square$

LEMMA 7.19. $\mathcal{A}(\emptyset) \implies 2 \cdot L(2)$.

PROOF. We begin with the amalgamation: [diagram].

If $2 \cdot L(2)$ is omitted in an amalgamation class containing $\mathcal{A}(\emptyset)$ then Lemma 7.18 together with the foregoing amalgamation provides the factors for both forms of the first amalgamation following, whose results can be used to form the second one; the latter suffices.

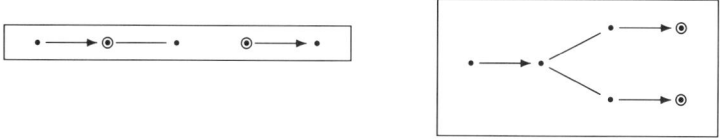

$\square$

LEMMA 7.20. $\mathcal{A}(\emptyset) \implies L(2) + P_3$.

PROOF. Suppose that H is a homogeneous directed graph embedding all graphs in $\mathcal{A}(\emptyset)$ and omitting $L(2) + P_3$. We consider oriented paths of length 4:

$$\cdot \xrightarrow{\alpha} \cdot \xrightarrow{\beta} \cdot \xrightarrow{\gamma} \cdot \xrightarrow{\delta} \cdot$$

where $\alpha, \beta, \gamma, \delta$ are orientations which will be encoded by binary digits 0 (rightward) and 1 (leftward). The relevant paths are those with the following codes:

(1) 0000 (2) 0011 (3) 1100;
(4) 0001 (5) 0010 (6) 0100 (7) 1000.

Step I. Either H embeds (1-3) and omits (4-7), or H omits (1-3) and embeds (4-7).

We use a variety of related amalgamations of the two forms shown:

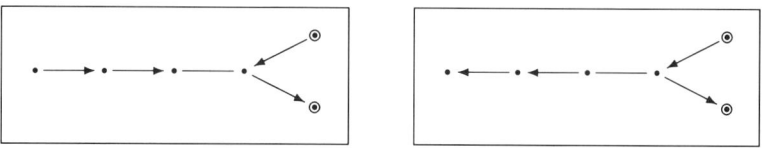

The relevant pairs of factors are the paths (1) and (4), (2) and (5), (3) and (6), or (1) and (7). H will omit at least one entry out of each of these pairs. On the other hand, Lemma 7.19 enables us to form the two amalgamations shown:

(I) [diagram] (II) [diagram]

from which we conclude that in each of the following pairs at least one entry embeds in H: (1), (5); (2), (4); (3), (7); (1), (6). From all of this it follows that H embeds (1-3) and omits (4-7), or vice versa, as claimed.

Step II. H omits (1-3) and embeds (4-7). Suppose on the contrary that H embeds (1-3) and omits (4-7). We use the following amalgamation, where the result of the first diagram will allow us to form either the second or the third:

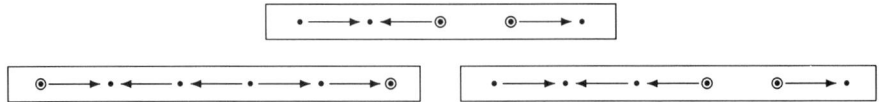

Step III. H embeds $L(2) + [I_1, I_2]$

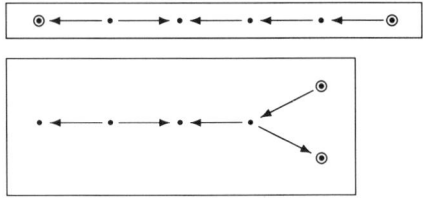

Step IV. Final construction

We aim at the following amalgamation diagram:

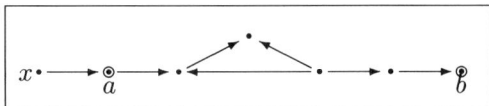

If $a \perp b$, $a \longrightarrow b$, or $b \longrightarrow a$, then we have embedded $L(2) + P_3$, (2), or (1) in H respectively. In all cases we have a contradiction.

To get the factor omitting b we use:

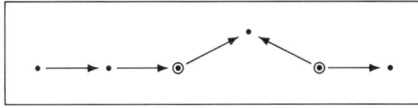

For the factor omitting a, we fix x in H and work in $x^\perp$. So we are aiming at:

(*)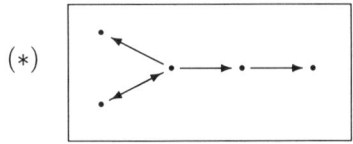

For this we use: $[\ \bullet \longrightarrow \odot \longleftarrow \bullet \longleftarrow \odot \longrightarrow \bullet \longrightarrow \bullet\ ]$.

$\square$

Now Proposition 25 reduces to the claim that $\mathcal{A}(\emptyset)$ entails $L(2) + P_3$ and $L(2) + I_n$ for all n. This follows from Lemmas 7.18 and 7.19.

7.7. Step 1, Proposition 26: the operations ±

We recall that the attachment operations A^+ and A^- attach oriented edges systematically to the points of A. Proposition 26 says that the set of consequences of $\mathcal{A}(\emptyset)$ is closed under these attachment operations. In our proof we shall make use of Proposition 25.

LEMMA 7.21. $\mathcal{A}(\emptyset) \implies [\cdot \longleftarrow \cdot \longrightarrow \cdot \longrightarrow \cdot]$.

PROOF.

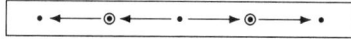

□

LEMMA 7.22. *Any amalgamation class of finite directed graphs containing $\mathcal{A}(\emptyset)$ contains either $[I_1, I_3]$ or the following directed graph Q_4:*

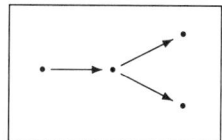

PROOF. Let H be a homogeneous directed graph embedding $I_1 + P_3$ and I_∞, and omitting both $[I_1, I_3]$ and Q_4. The following three amalgamation diagrams:

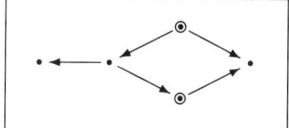 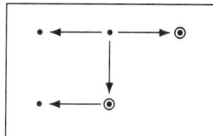 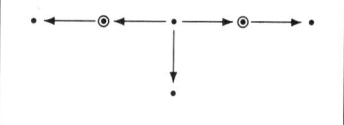

will produce a compatible pair of factors for the following:

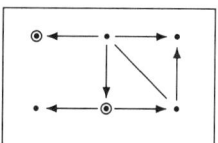

This one produces a copy of $[I_1, I_3]$ or Q_4 in all cases. □

LEMMA 7.23. $\mathcal{A}(\emptyset) \implies [I_1, I_3]$ *and* Q_4 *(in the notation of the previous lemma).*

PROOF. Let H be a homogeneous directed graph embedding $I_1 + P_3$ and I_∞. We first show that $[I_1, I_3]$ embeds in H, and by the previous lemma we may suppose Q_4 (and even $I_\infty + Q_4$) embeds in H for this part of the argument. The following amalgamations force $[I_1, I_3]$ into H (with a in one factor and b_1 and b_2 in the other):

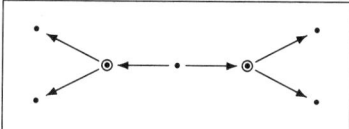 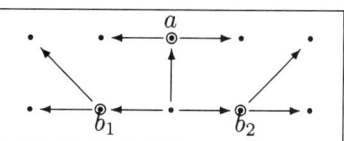

Lastly we must force Q_4 into H. Assume H omits Q_4. We shall show that H embeds:

(1)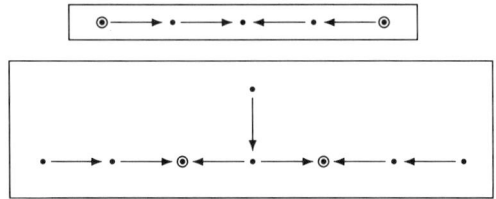

We use Lemma 7.21 and the two amalgamation diagrams following. In the first of these, no edge will be inserted, and we can then construct the second.

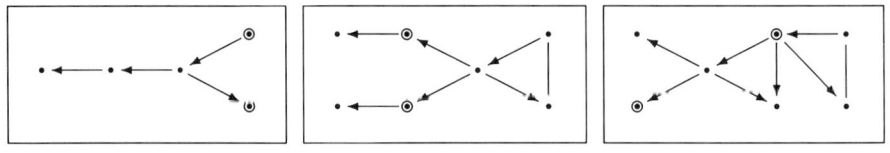

This yields (1).

We then perform the following three amalgamations in succession, forcing Q_4 into H:

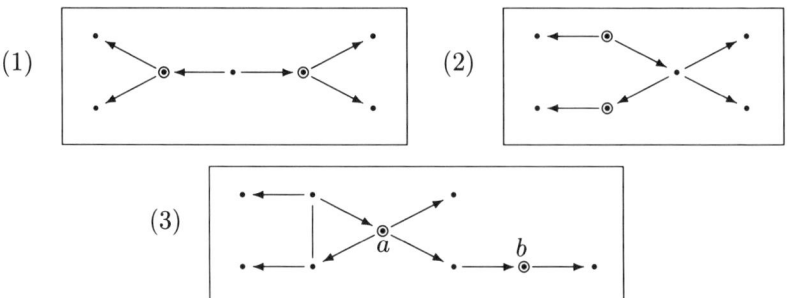

$\square$

LEMMA 7.24. $\mathcal{A}(\emptyset) \implies P_3^+$.

PROOF. We use the following amalgamations:

(1) (2)

(3)

We can set up (1) using the previous Lemma (and Lemma 7.18, which we have been using freely throughout). The result of (1) can be used to set up (2). If no edge is inserted, we obtain P_3^+; otherwise we obtain the factor of (3) omitting b. For the other factor of (3), it suffices to check that for A a copy of P_3 in H, $A^\perp$ embeds P_3 and hence without loss of generality embeds every element of $\mathcal{A}(\emptyset)$. This follows from the following amalgamation, which yields either $P_3 + P_3$ or P_3^+.

The factors can be obtained by applying Proposition 25. $\square$

Now we prove Proposition 26.

PROOF. It suffices to show that if A is a consequence of $\mathcal{A}(\emptyset)$, then A^+ is also a consequence of $\mathcal{A}(\emptyset)$. By Lemmas 7.5 and 7.18 we need only show that $\mathcal{A}(\emptyset) \implies I_n^+$ and $(I_1 + P_3)^+$. As $I_n^+ = n \cdot L(2)$ and $(I_1 + P_3)^+ = L(2) + P_3^+$, by Lemma 25 we need only check that $\mathcal{A}(\emptyset) \implies P_3^+$, which was done in the preceding lemma. □

7.8. Step 1, Propositions 27 and 28: some 1-types

In the present section $\mathbb{H}$ denotes an ample homogeneous 2-directed graph. We are concerned with the realization of 1-types over $2 \cdot L(2)$ and P_3. We may suppose that $\mathbb{H}$ realizes at least two cross types.

Recall that a 2-directed graph $\mathcal{A}$ is called $\mathbb{H}$-*constrained* if each of its restrictions to a 1-type over a tournament embeds in $\mathbb{H}$. A 2-type r realized in H_1 will be called a *ramsey type* for the subgraph H of H_2 if every finite $\mathbb{H}$-constrained 2-directed graph of the form (R, A) with R r-ramsey and $A \subseteq H$ embeds in $\mathbb{H}$.

LEMMA 7.25. *There is a ramsey type over I_∞.*

PROOF. As $\mathbb{H}$ is assumed ample this is a consequence of Lachlan's Ramsey argument. Much the same point was made in the proof of Lemma 7.13. □

LEMMA 7.26. *Every $\mathbb{H}$-constrained 1-type $\mathbb{A}$ over $I_1 + L(2)$ embeds in $\mathbb{H}$.*

PROOF. Let r be a ramsey type over I_∞. Let $R = \{x_1, x_2, x_3\}$ be r-ramsey. Let $B_0 = \{a, b\} \simeq I_2$, $B_i = B_0 \cup \{y_i\}$ for $i = 1, 2$ with $B_1 \simeq I_3$, $B_2 \simeq I_1 + L(2)$, and more specifically $b \longrightarrow y_2$. Determine $\text{tp}(x_1/y_1 b y_2)$ and $\text{tp}(x_2, x_3/ay_1 y_2)$ so that the amalgamation of (R, B_1) with (R, B_2) over (R, B_0) forces $\mathbb{A}$ into $\mathbb{H}$.

It remains to determine $\text{tp}(x_1/a)$ and $\text{tp}(x_2 x_3/b)$ so that the factors (R, B_1) and (R, B_2) embed in $\mathbb{H}$. As r is a ramsey type over B_1 we need only concern ourselves with (R, B_2). We first determine $\text{tp}(x_1/a)$ by amalgamating $(x_1, \{by_2\})$ with $(\emptyset, B_2)$ over $(\emptyset, \{by_2\})$, and we then form (R, B_2) by amalgamating $(R, \{ay_2\})$ with (x_1, B_2) over $(x_1, \{ay_2\})$. The factor $(R, \{ay_2\})$ is available since $\{ay_2\} \simeq I_2$. □

LEMMA 7.27. *If P is a 1-type over $L \simeq L(2)$ realized in $\mathbb{H}$ and $\mathbb{H}^P = ({}^P L, L^\perp)$, then $\mathbb{H}^P$ is ample.*

PROOF. This means that $\mathbb{H}$ embeds every $\mathbb{H}$-constrained 1-type over $L(2) + I_n$ for all n, which follows from Lemma 7.26 by induction on n, and that H_2 embeds $L(2) + P_3$, which follows from Proposition 25. □

LEMMA 7.28. *Let (x, L_i) $(i = 1, 2)$ be 1-types over $L_i \simeq L(2)$ which are realized in $\mathbb{H}$. Then there is a 1-type (x, L) over $L \simeq L(2)$ such that both of the extensions $(x, L_1 + L)$, $(x, L_2 + L)$ embed in $\mathbb{H}$.*

PROOF. We use the following amalgamation:

whose factors are afforded by Lemma 7.26 and Propositions 25 and 26. □

LEMMA 7.29. *Let $\mathbb{H}$ be an ample homogeneous 2-directed graph which embeds N 1-types over $L(2)$. Assume that Proposition 27 applies to every ample homogeneous 2-directed graph which embeds at most N 1-types over $L(2)$. Then there is a ramsey type for $\mathbb{H}$ over $\infty \cdot L(2)$.*

PROOF. By Lachlan's ramsey argument it suffices to show that every $\mathbb{H}$-constrained 1-type over $n \cdot L(2)$ (n arbitrary) embeds in $\mathbb{H}$. We prove this simultaneously for all ample homogeneous 2-directed graphs with at most N 1-types over $L(2)$, by induction on n. This is a direct application of Proposition 27 for this class of 2-directed graphs, using derived 2-directed graphs of the form $(^P L, L^{\perp})$ with $L \subseteq H_2$, $L \simeq L(2)$. □

We now place ourselves in the following situation until the completion of the proof of Proposition 27. $\mathbb{H}$ is an ample homogeneous 2-directed graph realizing exactly N 1-types over $L(2)$, and Proposition 27 is assumed to hold for all ample homogeneous 2-directed graphs realizing fewer than N 1-types over $L(2)$.

We shall use the following notation. If P, Q are 1-types over $L(2)$ then $P + Q$ denotes the 1-type over $2 \cdot L(2)$ whose restrictions to the components of $2 \cdot L(2)$ are P and Q.

LEMMA 7.30. *Let P, Q be 1-types over $L(2)$ which are realized in $\mathbb{H}$. If $Q + Q$ is realized in $\mathbb{H}$, then $P + Q$ is realized in $\mathbb{H}$.*

PROOF. Let $\mathbb{H}^P = (^P L, L^{\perp})$, $\mathbb{H}^Q = (^Q L, L^{\perp})$, where $L \subseteq H_2$, $L \simeq L(2)$. By Lemma 7.26 the cross types of $\mathbb{H}^P$ and $\mathbb{H}^Q$ coincide with those of $\mathbb{H}$. If $\mathbb{H}$ omits $P + Q$ then each of these 2-directed graphs realizes fewer than N 1-types over $L(2)$, and hence by our standing hypothesis Proposition 27 applies to these 2-directed graphs.

Let r be a ramsey type for $\mathbb{H}^Q$ over $\infty \cdot L(2)$ and let R be an r-ramsey graph on three vertices x_1, x_2, x_3. We shall now describe a particular amalgamation diagram which forces an embedding of $P + Q$ into $\mathbb{H}$.

Let $A_0 = I + L$ with $I = \{a_1, a_2\} \simeq I_2$, and with $L \simeq L(2)$. For $i = 1, 2$ let $A_i = A_0 \cup \{b_i\}$ with $a_i \longrightarrow b_i$. We shall construct factors (R, A_1) and (R, A_2) in $\mathbb{H}$ agreeing on (R, A_0), in such a way that in their amalgam one of the following will necessarily be isomorphic to $P + Q$:

$$(x_1, I \cup \{b_1, b_2\}); \quad (x_2, \{b_1, b_2\} \cup L); \quad (x_3, \{b_1, b_2\} \cup L);$$

according as $b_1 \perp b_2$, $b_1 \longrightarrow b_2$, or $b_2 \longrightarrow b_1$. In other words we require (in an obvious sense):

$$\text{tp}(x_1/a_1 b_1) = \text{tp}(x_2/b_1 b_2) = \text{tp}(x_3/b_2 b_1) = P;$$

$$\text{tp}(x_1/a_2 b_2) = \text{tp}(x_2/L) = \text{tp}(x_3/L) = Q.$$

We also fix a 1-type Q^* over $L(2)$ realized in both $\mathbb{H}^P$ and in $\mathbb{H}^Q$ (by Lemma 7.28) and we require $\text{tp}(x_1/L) = Q^*$.

It remains to determine $\text{tp}(x_2 x_3/a_1 a_2)$ so that both factors embed in $\mathbb{H}$. One obvious constraint is easily met:

$$\text{tp}(x_j/a_2 b_2) \text{ is realized in } \mathbb{H}^Q \text{ for } j = 2, 3.$$

If $\text{tp}(x_2 x_3/a_2)$ is chosen so that this constraint is met then the factor (R, A_2) will embed in $\mathbb{H}^Q$ and *a fortiori* in $\mathbb{H}$ for any choice of $\text{tp}(x_2 x_3/a_1)$, since the 1-types over $L(2)$ involved in (R, A_2) are all realized in $\mathbb{H}^Q$.

It remains to determine $\text{tp}(x_2 x_3/a_1)$ so that (R, A_1) embeds in $\mathbb{H}$, and for this we simply amalgamate the subfactors $(R, \{b_1, a_2\} \cup L)$ and (x_1, A_1), working over $(x_1, \{b_1, a_2\} \cup L)$. The first of these subfactors already embeds in $\mathbb{H}^Q$ since r is a ramsey type for H^Q, and the second is of the form $(x, I_1 + 2 \cdot L(2))$ with

7.8. STEP 1, PROPOSITIONS 27 AND 28: SOME 1-TYPES   135

$x, 2 \cdot L(2)) \simeq P + Q^*$, which may be seen to embed in $\mathbb{H}$ by applying Lemma 7.19 to $\mathbb{H}^P$. □

LEMMA 7.31. *Let P be a 1-type over $L(2)$ such that $P + P$ is not realized in $\mathbb{H}$. Then there is a cross type p such that for every n, for every 1-type Q_1 over I_n, and for every 1-type Q_2 over $L = \{a, b\}$ with $a \longrightarrow b$ whose restriction to a is p, the 1-type $P + Q_1 + Q_2$ is realized in $\mathbb{H}$.*

PROOF. Suppose on the contrary that for each cross type p we have a counterexample n_p, $Q_1(p)$, $Q_2(p)$. Let $Q = \oplus_p Q_1(p)$, $n = \sum_p n(p)$. Let k be the number of cross types in $\mathbb{H}$. Amalgamate $(x, I_n + I_k + L(2))$ with $(\emptyset, [\{y\}, I_k] + I_n + L(2))$ where $\text{tp}(x/I_n) = Q$, $\text{tp}(x/L(2)) = P$, and $\text{tp}(x/I_k)$ is made up of all the cross types. The second factor embeds in $\mathbb{H}$ since H_2 embeds all elements of $\mathcal{A}(\emptyset)$. Take $p = \text{tp}(xy)$ in the amalgam to arrive at a contradiction. □

LEMMA 7.32. *Let P be a 1-type over $L(2)$ such that $P + P$ is not realized in $\mathbb{H}$ and let r be a ramsey 2-type for $\mathbb{H}^P$ over I_∞, p a cross type. There is a 1-type Q over $L(2) = [a \longrightarrow b]$ whose restriction to a is p, such that for any finite configuration $\mathbb{A} = (R, I_n)$ with R r-ramsey, and any $x_0 \in R$, $\mathbb{H}$ embeds $(R, I_n + L)$ with:*

$$(R, I_n) \simeq \mathbb{A}, \quad (x_0, L) \simeq Q, \quad (x, L) \simeq P \text{ for all } x \in R - \{x_0\}.$$

PROOF. Suppose on the contrary that for each cross type q the type Q represented by $[p \longrightarrow q]$ does not have the desired property, so that we have counterexamples $R(q), I_{n(q)}, x_0(q)$. Take (R, I_n, x_0) isomorphically embedding each configuration $(R(q), I_{n(q)}, x_0(q))$. To get a contradiction it suffices to amalgamate $(R, I_n + \{a\})$ with $(R - \{x_0\}, I_n + L)$, where $L = [a \longrightarrow b]$. Here $(R - \{x_0\}, I_n + L)$ embeds in $\mathbb{H}$ since $(R - \{x_0\}, I_n)$ embeds in $\mathbb{H}^P$. □

Now we shall prove Proposition 27.

PROOF. A counterexample consists of an ample homogeneous 2-directed graph $\mathbb{H}$ which fails to embed $P + P^*$ for some choice of 1-types P, P^* over $L(2)$ which do embed in $\mathbb{H}$. Let N be the number of 1-types over $L(2)$ realized in $\mathbb{H}$, which we take to be minimized over all counterexamples. By Lemma 7.30, we may suppose that the 1-type $P + P$ is omitted by $\mathbb{H}$. Let P be of the form $p_1 \longrightarrow p_2$, let r be ramsey for $\mathbb{H}^P$ over I_∞, and let R be r-ramsey of order 3, $R = \{x_0, x_1, x_2\}$.

Let $A_0 = \{b_1, b_2\} + L \simeq I_2 + L(2)$, $A_i = A_0 \cup \{a_i\}$ for $i = 1, 2$ with $a_i \longrightarrow b_i$. It suffices to amalgamate (R, A_1) with (R, A_2), where:

$$\begin{aligned}
\text{tp}(x_0/a_i b_i) &= P & \text{for} & \quad i = 1, 2; \\
\text{tp}(x_j/L) &= P & \text{for} & \quad j > 0; \\
\text{tp}(x_i/a_i) &= p_1 & \text{for} & \quad i = 1, 2; \\
\text{tp}(x_j/a_i) &= p_2 & \text{for} & \quad i + j = 3.
\end{aligned}$$

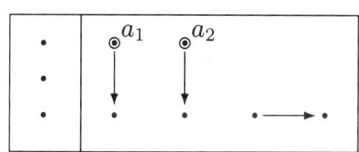

So our problem is to find compatible forms of these two factors in $\mathbb{H}$. We choose a cross type p according to Lemma 7.31 and then we choose a 1-type Q over $L(2)$

according to Lemma 7.32, and we shall require in addition:
$$\operatorname{tp}(x_0/L) = Q.$$

Our construction will be based on the following claim:

(*) There are cross types s_1, s_2 such that for any n and any configuration $\mathbb{A}$ of the form (R, I_n), the configuration $(R, I_n + L_1 + L_2)$ embeds in $\mathbb{H}$, where $(R, I_n) \simeq \mathbb{A}$, $L_i \simeq L(2)$, and:
$$\operatorname{tp}(x_0/L_1 + L_2) = Q + P;$$
$$\operatorname{tp}(x_i/L_1 + L_2) = P + [p_i \longrightarrow s_i] \quad \text{for} \quad i = 1, 2.$$

We make a similar claim (**) concerning the analogous situation with p_1, p_2 interchanged; call the corresponding cross types t_1, t_2. The definition of (R, A_1) and (R, A_2) can be completed on the basis of these two claims by taking:

$$\operatorname{tp}(x_i/b_1) = s_i \text{ for } i = 1, 2;$$

$$\operatorname{tp}(x_i/b_2) = t_i \text{ for } i = 1, 2;$$

So it suffices to check (*), as the treatment of (**) is entirely analogous. We suppose accordingly that for all choices of cross types s_1, s_2 there are counterexamples $n(s_1, s_2)$, $(R, I_{n(s_1, s_2)})$, and $x_0(s_1, s_2)$ to (*). Embed all of these into a configuration (R, I_n, x_0). Then to reach a contradiction it will be sufficient to amalgamate $(R, I_n + L_1 + \{a\})$ with a configuration $(\{x_0\}, I_n + L_1 + L_2)$ where $L_2 = [a \longrightarrow b]$ and:

$$\operatorname{tp}(x_0/L_1 + L_2) = Q + P;$$

$$\operatorname{tp}(x_i/L_1 + (a)) = P + p_i \text{ for } i = 1, 2.$$

The factor $(R, I_n + L_1 + \{a\})$ embeds in $\mathbb{H}$ by the choice of Q, and the factor $(x_0, I_n + L_1 + L_2)$ embeds in $\mathbb{H}$ by the choice of p.

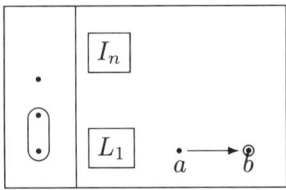

$\square$

Now we can prove Proposition 28 quickly. The assertion is that an ample homogeneous 2-directed graph $\mathbb{H}$ will embed all $\mathbb{H}$-constrained 1-types over P_3.

PROOF. By Lemma 7.29 there is a ramsey 2-type r over $\infty \cdot L(2)$ relative to $\mathbb{H}$. Let R be r-ramsey of order 3, $R = \{x_1, x_2, x_3\}$. Let $A_0 = \{a_1, a_2, a_3\} \simeq I_3$, $A_i = A_0 \cup \{b_i\}$, with $a_1 \longrightarrow b_1 \longrightarrow a_2 a_3$ and $a_3 \longrightarrow b_2$. Choose $\operatorname{tp}(x_1/b_1 a_3 b_2)$, $\operatorname{tp}(x_2/b_2 b_1 a_2)$, and $\operatorname{tp}(x_3/a_1 b_1 b_2)$ so that after amalgamating (R, A_1) and (R, A_2) over (R, A_0), one of these will be the desired 1-type over P_3. As $A_2 \simeq I_2 + L(2)$, it suffices to embed (R, A_1) in $\mathbb{H}$ in such a way that each $\operatorname{tp}(x_i/a_3 b_2)$ is realized in $\mathbb{H}$. This is already the case for $i = 1$, but $\operatorname{tp}(x_2 x_3/a_3)$ remains to be chosen.

7.8. STEP 1, PROPOSITIONS 27 AND 28: SOME 1-TYPES 137

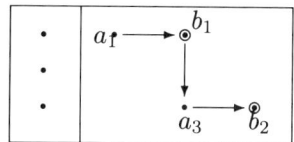

Form $A = A_1 \cup A_2$ with no new edges. Amalgamate:

$$(\emptyset, A) \text{ and } (x_3, \{a_1 b_1 b_2\}) \text{ over } (\emptyset, \{a_1 b_1 b_2\})$$

to form (x_3, A) in $\mathbb{H}$. Amalgamate $(\{x_3\}, A)$ with $(\{x_2 x_3\}, \{b_1 a_2 b_2\})$ over the common part $(x_3, \{b_1 a_2 b_2\})$ to form $(\{x_2, x_3\}, A)$ embedded in $\mathbb{H}$. In particular $\mathrm{tp}(x_i/a_3 b_2)$ is realized in $\mathbb{H}$ for $i = 2, 3$. Now amalgamate $(R, \{b_1 a_3\})$ with $(\{x_2 x_3\}, A_1)$ over $(\{x_2 x_3\}, \{b_1 a_3\})$ to form (R, A_1) of the desired type. □

CHAPTER 8

Theorems 7.6-7.9

To complete the proof of the classification theorem for primitive homogeneous directed graphs embedding I_∞, it remains to carry out steps 2, 3, 4, 6 of our outline, dealing with the derivation of Theorems 7.6–7.9 under appropriate inductive hypotheses.

8.1. Step 2. Theorems 7.6 and 7.7

In the present section $\mathcal{T}$ is a fixed finite set of finite tournaments and we assume Proposition 26, and Theorem 7.7.$\mathcal{T}_0$ and MT 3.$\mathcal{T}_0$ for $\mathcal{T}_0 < \mathcal{T}$. Proposition 26 contains Proposition 25. We shall prove Theorems 7.6 and 7.7 for $\mathcal{T}$:

Theorem 7.6: Suppose that $T \in \mathcal{T}$, A is a finite $\mathcal{T}$-constrained directed graph containing T, $b \in A$, and every tournament contained in $A^* = A - \{b\}$ is smaller than T. Then $\mathcal{A}(\mathcal{T}) \implies A$.

Theorem 7.7: If $\{a, b\} \cup T$ is $\mathcal{T}$-constrained with T a tournament and $a \perp T$, then $\mathcal{A}(\mathcal{T}) \implies \{a, b\} \cup T$.

LEMMA 8.1. $\mathcal{A}(\mathcal{T}) \implies I_1 + \mathcal{A}(\mathcal{T})$

PROOF. That $\mathcal{A}(\emptyset) \implies I_1 + \mathcal{A}(\emptyset)$ is contained in Proposition 25. It remains to show that $\mathcal{A}(\mathcal{T}) \implies I_1 + T$ for $T \in \mathcal{T}$. We may suppose that $|T| > 2$.

Let H be a homogeneous directed graph embedding all graphs in $\mathcal{A}(\mathcal{T})$, fix $a, b \in T$, let $T^* = T - \{a, b\}$, and let U_1, U_2, U_3 be disjoint copies of T^*. Let $A_0 = (U_1 + U_2 + U_3) \cup \{c_3, d\}$ with $d \perp U_1 + U_2 + U_3$, $(\{c_3\}, U_3) \simeq (b, T^*)$, and $\text{tp}(c_3/U_1 + U_2 + \{d\})$ not yet specified. Set $A_i = A_0 \cup \{c_i\}$ for $i = 1, 2$ where:

$$(c_1, U_1) \simeq (a, T^*); \qquad (c_1, U_2) \simeq (b, T^*);$$
$$(c_2, U_1) \simeq (b, T^*); \qquad (c_2, U_2) \simeq (a, T^*);$$
$$c_1 \perp U_3 \cup \{c_3, d\}; \quad (c_2, c_3, U_3) \simeq (a, b, T^*); \quad c_2 \perp d.$$

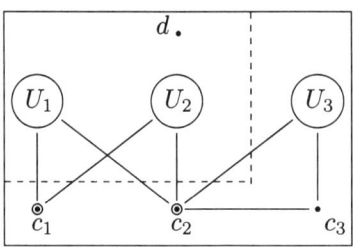

If suitable factors A_1, A_2 embed in H, then their amalgamation will produce the desired copy of $I_1 + T$, with c_1 or d representing I_1.

The factor A_2 may be produced by amalgamating $(U_1 + U_2 + U_3) \cup \{c_2, d\}$ with $U_3 \cup \{c_2, c_3, d\}$ (for a suitable choice of $\text{tp}(c_3 d)$) over $U_3 \cup \{c_2, d\}$. The first subfactor embeds in H by an application of MT 3.$\mathcal{T}_0$ with $\mathcal{T}_0$ the set of its subtournaments. Some form of the second subfactor is obtained by amalgamating $U_3 \cup \{c_2, c_3\} \simeq T$ with $U_3 \cup \{c_2, d\}$.

The factor A_1 corresponding to A_2 is now completely determined. Let $\mathcal{T}_1$ be the set of its subtournaments. As they are all smaller than $|T|$, $\mathcal{T}_1 < \mathcal{T}$ and hence by MT 3.$\mathcal{T}_1$ it will suffice to check that these tournaments all embed in H. In fact they are all contained either in A_2, or in a copy of T. □

LEMMA 8.2. *If $T \in \mathcal{T}$, $b \in T$, $a \longrightarrow b$, and $a \perp T - \{b\}$, then $\mathcal{A}(\mathcal{T}) \implies \{a\} \cup T$.*

PROOF. If $|T| \leq 2$ Proposition 26 applies. Now assume that $b, c, d \in T$ are distinct with $c \longrightarrow d$. Let $T_1 = T - \{c, d\}$, and let T_2, T_3 be two more disjoint copies of T_1. Let $b_i \in T_i$ for $i = 2, 3$ correspond to $b \in T_1$. Let $S_0 = (T_1 + T_2 + T_3) \cup \{a_2, a_3, c\}$ where $a_2 \longrightarrow b_2$, $a_3 \longrightarrow b_3$, $(c, T_2) \simeq (c, T_1)$, $(c, T_3) \simeq (d, T_1)$ and there are no further new edges.

The collection $\mathcal{T}_0$ of subtournaments of S_0 satisfies $\mathrm{rk}\mathcal{T}_0 < \mathrm{rk}\mathcal{T}$, so $\mathcal{T}_0 < \mathcal{T}$ and $\mathcal{A}(\mathcal{T}_0) \implies S_0$, hence $\mathcal{A}(\mathcal{T}) \implies S_0$.

Let $\mathcal{A} \supseteq \mathcal{A}(\mathcal{T})$ be an amalgamation class. Then $T + I_2 \in \mathcal{A}$ by Lemma 8.1, so $\mathcal{A}$ contains an amalgam A_1 of $T + \{a_2, a_3\}$ with S_0 over $T_1 \cup \{a_2, a_3, c\}$ (recall $T_1 \cup \{c\} \subseteq T$). Let $A_0 = A_1 - \{c\}$, $A_2 = A_0 \cup \{a_1\}$ with $a_1 \longrightarrow b$, and $(a_1, T_2) \simeq (d, T_1)$, $(a_1, T_3) \simeq (c, T_1)$, $a_1 \perp T_1 - \{b\}, d$. Amalgamation of A_1 with A_2 over A_0 forces $\{a\} \cup T$ into $\mathcal{A}$:

$$\begin{array}{lll} \text{if} & a_1 \perp c: & \{a_1\} \cup T \simeq \{a\} \cup T; \\ \text{if} & a_1 \longrightarrow c: & \{a_3\} \cup (T_3 \cup \{a_1 c\}) \simeq \{a\} \cup T; \\ \text{if} & a_1 \longleftarrow c: & \{a_2\} \cup (T_2 \cup \{c a_1\}) \simeq \{a\} \cup T; \end{array}$$

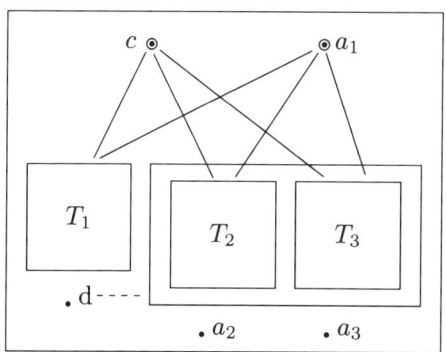

It remains to be checked that A_2 is in $\mathcal{A}$. Now:

$$A_2 = (T_1 + T_2 + T_3) \cup \{a_1, a_2, a_3, d\} \text{ with } a_1 \perp d$$

and a_2 and a_3 are not contained in any tournaments of order greater than 2. Hence the set $\mathcal{T}_1$ of subtournaments of A_2 satisfies $\mathcal{T}_1 < \mathcal{T}$, and by induction $\mathcal{A}(\mathcal{T}_1) \implies A_2$. We must check that $\mathcal{A}(\mathcal{T}_1) \subseteq \mathcal{A}$. A tournament $T \in \mathcal{T}_1$ either omits a_1, and lies in A_0, or contains a_1 and lies in some $\{a_1\} \cup T_i$, hence in $\mathcal{A}(\mathcal{T})$. In either case $T \in \mathcal{A}$. □

We now prove Theorem 7.6.$\mathcal{T}$:

Suppose that $T \in \mathcal{T}$, A is a finite $\mathcal{T}$-constrained directed graph containing T, $b \in A$, and every tournament contained in $A^* = A - \{b\}$ is smaller than T. Then $\mathcal{A}(\mathcal{T}) \implies A$.

PROOF. If $|T| \leq 2$ then Proposition 26 applies. Assume $|T| > 2$. Let H be a homogeneous directed graph embedding all graphs in $\mathcal{A}(\mathcal{T})$. Let $a \in H$, and $\mathbb{H} = (a', a^\perp)$. Then $\mathbb{H}$ is ample. It suffices to show that (b, A^*) embeds in $\mathbb{H}$. Let $\mathcal{T}_0$ be the set of subtournaments of A^*. By MT 3.$\mathcal{T}_0$ it suffices to show that (b, T_0) embeds in $\mathbb{H}$ for $T_0 \in \mathcal{T}_0$.

Thus we claim that for T_0 a subtournament of $A-\{b\}$, the graph $\{a,b\}\cup T_0$ with $a \longrightarrow b$ and $a \perp T_0$ embeds in H. Let $\mathcal{T}_1$ be the set of subtournaments of $\{a,b\}\cup T_0$. If $\mathcal{T}_1 < \mathcal{T}$ then MT 3.$\mathcal{T}_1$ yields our claim, and otherwise we have $\{b\}\cup T_0 \in \mathcal{T}$, and Lemma 8.2 applies. □

We shall prove Theorem 7.7.$\mathcal{T}$:

If $\{a,b\}\cup T$ is $\mathcal{T}$-constrained with T a tournament and $a \perp T$, then $\mathcal{A}(\mathcal{T}) \implies \{a,b\}\cup T$.

PROOF. With $\mathcal{T}$ fixed we proceed by induction on $|T|$ (in addition to our standard inductive framework).

If $a \perp b$ then by Lemma 8.1 it suffices to show $\mathcal{A} \implies \{b\}\cup T$. As a result, we only need to consider one case explicitly: $a \longrightarrow b$.

The preceding lemma covers the case in which $\{b\}\cup T$ is a tournament. The case in which $b \perp T - \{b\}$ is covered by Theorem 7.6. Accordingly we may assume there are $c, d \in T$ with $b \perp c$, $b \not\perp d$. For definiteness we suppose that $b \longrightarrow d$; the orientation is of little significance.

Let T_1, T_2, T_3 be disjoint copies of $T - \{c\}$, $T - \{d\}$, $T - \{d\}$ respectively, and let:

$$A_0 = (T_1 + T_2 + T_3) + \{a_1, a_2\} \simeq T_1 + T_2 + T_3 + I_2;$$

$$A_i = A_0 \cup \{b_i\}; \quad a_i \longrightarrow b_i \text{ and:}$$

$$(b_1, T_1) \simeq (b, T - \{c\}); \quad (b_2, T_1) \simeq (c, T - \{c\});$$
$$(b_1, T_2) \simeq (b, T - \{d\}); \quad (b_2, T_2) \simeq (d, T - \{d\});$$
$$(b_1, T_3) \simeq (d, T - \{d\}); \quad (b_2, T_3) \simeq (b, T - \{d\}).$$

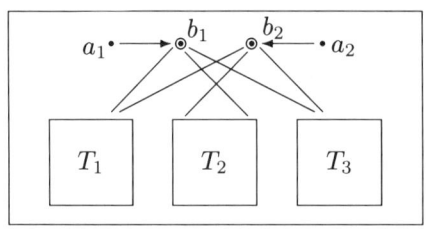

Any amalgam of A_1, A_2 over A_0 will contain a copy of $\{a,b\}\cup T$. It therefore suffices to check that A_1, A_2 are consequences of $\mathcal{A}(\mathcal{T})$. Let $A_1^* = A_1 - \{a_2\}$, $A_2^* = A_2 - \{a_1\}$. Then $A_i \simeq A_i^* + I_1$, so it suffices to check that $\mathcal{A}(\mathcal{T}) \implies A_i^*$ for $i = 1, 2$.

To see this, we shall pass to the 2-directed graph $\mathbb{H} = (a_i', a_i^\perp)$. It suffices to check that $\mathbb{H}$ embeds $(\{b_i\}, T_1 + T_2 + T_3)$. This 2-directed graph is ample as a consequence of Proposition 26, applied repeatedly to I_1. Let $\mathcal{T}_0$ be the set of subtournaments of $T_1 + T_2 + T_3$. Then $\mathcal{T}_0 < \mathcal{T}$. By MT 3.$\mathcal{T}_0$ it suffices to check that $\mathbb{H}$ embeds $(\{b_i\}, T_j)$ for $j = 1, 2, 3$. Let $\mathcal{T}_j$ be the set of finite subtournaments of $\{a_i, b_i\} \cup T_j$. Then $\mathcal{T}_j \subseteq \mathcal{T}$ and hence we may apply Theorem 7.7.$\mathcal{T}_j$ or induction on $|T|$, as appropriate, getting an embedding of $(\{b_i\}, T_j)$ into $\mathbb{H}$. □

8.2. Step 5. Theorem 7.9.$\mathcal{T}$: extending a direct sum

The theorem asserts that $\mathcal{A}(\mathcal{T})$ entails any $\mathcal{T}$-constrained finite directed graph that is a one point extension of a disjoint sum of graphs in $\mathcal{A}(\mathcal{T})$.

In our proof we shall assume Propositions 26, MT 2.$\mathcal{T}$, and Theorem 7.7.$\mathcal{T}$. Proposition 25 is contained Proposition 26.

PROOF. Fix H a homogeneous directed graph embedding all graphs in $\mathcal{A}(\mathcal{T})$, and a graph $A = \oplus A_i$ with $A_i \in \mathcal{T} \cup \{P_3\}$, as well as a $\mathcal{T}$-constrained extension of A by one point x. Fix $a \in H$ and let $\mathbb{H} = (a', a^{\perp})$. This is ample by repeated applications of Proposition 26, and we shall show that (x, A) embeds in $\mathbb{H}$. By MT 2.$\mathcal{T}$ it suffices to show that (x, A_i) embeds in $\mathbb{H}$ for each i. For $A_i \in \mathcal{T}$ this is Theorem 7.7.$\mathcal{T}$. For $A_i = P_3$ this reduces by MT 2.$\mathcal{T}$ to the case $A_i = L(2)$ which we may suppose is in $\mathcal{T}$.

□

8.3. Step 3. Theorem 7.8, 1-types over sums

Our hypothesis throughout this section and the next is that Propositions 26 and 28 are known, as well as Theorems 7.6.$\mathcal{T}$ and 7.7.$\mathcal{T}$, and MT 3.$\mathcal{T}_0$ for $\mathcal{T}_0 < \mathcal{T}$. We derive Theorem 7.8.$\mathcal{T}$, concerning the embedding of 1-types over directed graphs which are disjoint sums of two tournaments. This is the final step in the proof of the classification theorem, and requires detailed amalgamation arguments which however have a more general flavor than most.

We also mention at the outset that we shall find it necessary to carry out induction over some additional parameters, as has happened on occasion earlier. We shall go into this at the beginning of the next section. In the proof of Theorem 7.8 we shall make a case division into three quite manageable cases and one more demanding case. The present section deals with the three easy cases, leaving the more difficult case for the next section.

We now fix an ample homogeneous 2-directed graph $\mathbb{H}$, tournaments T_1, T_2 in $\mathcal{T}$ and an H-constrained 1-type $\mathbb{A}$ over $T_1 + T_2$. Our goal of course is to embed $\mathbb{A}$ in $\mathbb{H}$. If $T_1 \not\simeq T_2$ we can proceed in a fairly direct manner after making a suitable case division. When $T_1 \simeq T_2$ we have already encountered the type of argument needed in the proof of Proposition 7.19, which dealt with the possibility $T_1 \simeq T_2 \simeq L(2)$. A generalization of this argument will be given in the next section.

Suppose toward a contradiction that $\mathbb{A}$ does not embed in $\mathbb{H}$. Taking $\mathcal{T}$ minimal, we conclude that $\mathcal{T}$ consists exactly of the subtournaments of T_1 and T_2 and that $\max(|T_1|, |T_2|)$ has been minimized. We shall assume as a matter of notation that $|T_2| \leq |T_1|$, and we shall assume that our counterexample has also been chosen to minimize $|T_2|$. Finally, in the very important case in which $T_1 \simeq T_2$, so that $\mathcal{T}$ consists of the subtournaments of a single tournament T, we shall assume that the example has also been chosen to minimize the number of 1-types realized in $\mathbb{H}$ over T.

For the present we deal with the following cases, which cover all possibilities in which T_1, T_2 are not isomorphic.

Case 1. $T_1 \not\simeq T_2$, and there is a tournament T with $|T| \leq |T_1|$ such that some 1-type Q over T is not realized in $\mathbb{H}$.

Case 2. $|T_2| < |T_1|$, and for all tournaments T with $|T| \leq |T_1|$ and for all 1-types Q over T, Q is realized in $\mathbb{H}$.

Case 3. $|T_1| = |T_2|$, T_1 is not isomorphic to T_2, and for all tournaments T with $|T| \leq |T_1|$ and all 1-types Q over T, Q is realized in $\mathbb{H}$.

Case 1. $T_1 \not\simeq T_2$, and there is a tournament T with $|T| \leq |T_1|$ such that some 1-type Q over T is not realized in $\mathbb{H}$.

DEFINITION 7. Fix Q, T as described with $|T|$ minimal. Fix $a_1 \in T_1$, $a_2 \in T_2$, $b_1, b_2 \in T$, and let T_1^*, T_2^*, T^* be $T_1 - \{a_1\}$, $T_2 - \{a_2\}$, and $T - \{b_1, b_2\}$ respectively. Let U_1, U_2 be disjoint copies of T^*, $A_0 = T_1^* + T_2^* + U_1 + U_2$, $A_i = A_0 \cup \{c_i\}$ for $i = 1, 2$ with:

$$(c_i, T_i^*) \simeq (a_i, T_i^*) \text{ for } i = 1, 2;$$
$$c_1 \perp T_2^*, \quad c_2 \perp T_1^*; \quad (c_i, U_i) \simeq (b_1, T^*) \quad \text{for } i = 1, 2$$
$$(c_1, U_2) \simeq (c_2, U_1) \simeq (b_2, T^*).$$

Let $\mathcal{T}_0$ be a finite set of finite tournaments closed under subtournament and representing exactly the isomorphism types of tournaments T_0 with $|T_0| < |T_1|$ or $T_0 \simeq T_2$ (or both). Applying Theorem 3.$\mathcal{T}_0$, let r be a ramsey type for $\mathbb{H}$ over the $\mathcal{A}(\mathcal{T}_0)$-generic graph. Let R be r-ramsey on three vertices x_1, x_2, x_3. We intend to amalgamate (R, A_1) with (R, A_2) over (R, A_0) in such a way that one of the following will be forced to occur:

$$(x_i, U_i \cup \{c_1, c_2\}) \simeq Q \text{ for some } i = 1 \text{ or } 2;$$

$$(x_3, (T_1^* + T_2^*) \cup \{c_1, c_2\}) \simeq \mathbb{A}.$$

It remains to construct suitable factors (R, A_1) and (R, A_2) embedding in $\mathbb{H}$. The former is obtained by amalgamating:

$$(x_3, A_1) \text{ with } (R, (U_1 + U_2 \cup \{c_1\}) + T_2^*).$$

The subfactor (x_3, A_1) is produced by amalgamating $(x_3, \{c_1\} + T_1^* + T_2^*)$ with $(\emptyset, A_1)$ over $(\emptyset, \{c_1\} + T_1^* + T_2^*)$; $(\emptyset, A_1)$ is afforded by Theorem 7.6. The second subfactor is afforded by the choice of R, if it is $\mathbb{H}$-constrained. In fact we shall constrain $\text{tp}(x_i/T_2^*)$ for $i = 1, 2$ by requiring it to be the restriction of a type $\text{tp}(x_i/\{c_2\} \cup T_2^*)$ realized in $\mathbb{H}$. Since $|U_i \cup \{c_1\}| < |T|$, it is not necessary to impose additional constraints on $\text{tp}(x_3/U_1 + U_2)$.

Now in view of the choice of R, to embed the corresponding factor (R, A_2) into $\mathbb{H}$ it suffices to check that it is $\mathbb{H}$-constrained, and this is immediate.

Case 2. $|T_2| < |T_1|$, and for all tournaments T with $|T| \leq |T_1|$ and for all 1-types Q over T, Q is realized in $\mathbb{H}$.

DEFINITION 8. Let r and R be as in Case 1. Fix $a, b \in T_1$ and $c \in T_2$, and let T_1^*, T_2^* be $T_1 - \{a, b\}$ and $T_2 - \{c\}$ respectively. Let U_1, U_2, U_3 be disjoint copies of

8.3. STEP 3. THEOREM 7.8, 1-TYPES OVER SUMS

T_1^* and let V_1, V_2 be disjoint copies of T_2^*. Let $A_0 = (\oplus_{i \leq 3} U_i + \oplus_{i \leq 2} V_i) \cup \{b_3, c_2\}$, where:

$$(b_3, U_3) \simeq (b, T_1^*); \quad (c_2, V_2) \simeq (c, T_2^*).$$

The type of b_3 over $U_1 + U_2 + \{c_2\}$ is still to be determined, but apart from this there will be no additional edges.

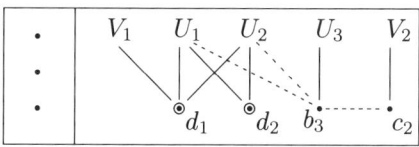

Let $A_i = A_0 \cup \{d_i\}$ for $i = 1, 2$ with:

$$(d_i, U_i) \simeq (a, T_1^*); \quad (d_1, U_2) \simeq (d_2, U_1) \simeq (b, T_1^*);$$

$$(d_1, V_1) \simeq (c, T_1^*); \quad (d_2, b_3, U_3) \simeq (a, b, T_1^*);$$

$$d_1, d_2 \perp c_2, V_2; \quad d_2 \perp V_1.$$

We shall amalgamate (R, A_1) with (R, A_2) over (R, A_0), choosing types so that one of the following will provide a copy of $\mathbb{A}$:

$$(x_i, (U_i \cup \{d_1, d_2\}) + (V_2 \cup \{c_2\})) \text{ for } i = 1 \text{ or } 2;$$

$$(x_3, [U_3 \cup \{b_3, d_2\}] + [V_1 \cup \{d_1\}]).$$

It remains to find suitable factors (R, A_1) and (R, A_2) embedding in $\mathbb{H}$. Once (R, A_2) has been determined it will be easy to find a matching factor (R, A_1) in $\mathbb{H}$, since the subtournaments of A_1 are all smaller than T_1.

Thus it suffices to construct a suitable factor (R, A_2) in $\mathbb{H}$. We amalgamate:

$$(R, [\oplus_{i \leq 3} U_i \cup \{d_2\}] + V_1 + [V_2 \cup \{c_2\}])$$
$$\text{with}$$
$$(R, V_1 + (U_3 \cup V_2 \cup \{d_2, b_3, c_2\})).$$

The former embeds in $\mathbb{H}$ by the choice of R. For the latter we amalgamate $(R, V_1 + (V_2 \cup \{d_2, b_3, c_2\}))$, which is afforded by the choice of R, with:

(∗) $\qquad (x_3, V_1 + (U_3 \cup V_2 \cup \{d_2, b_3, c_2\}));$

here the type of $b_3 c_2$ remains to be specified. So the final subfactor (∗) may be constructed by amalgamating $(x_3, V_1 + V_2 + (U_3 \cup \{d_2, b_3\}))$ with $(x_3, V_1 + (V_2 \cup \{c_2\}) + (U_3 \cup \{d_2\}))$. We shall check that these two configurations occur in $\mathbb{H}$.

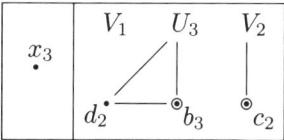

The configuration $(x_3, V_1 + (V_2 \cup \{c_2\}) + (U_3 \cup \{d_2\}))$ is afforded by induction on $\mathcal{T}$. To see that $(x_3, V_1 + V_2 + (U_3 \cup \{d_2, b_3\}))$ also occurs, let $p = \text{tp}(x_3/V_1)$, $\mathbb{H}^p = ({}^p V_1, V_1^\perp)$, and work in $\mathbb{H}^p$. By the minimality of the counterexample $\mathbb{H}, \mathbb{A}$, we have both $(x_3, U_3 \cup \{d_2, b_3\})$ and (x_3, V_2) in $\mathbb{H}^p$, and hence (again by minimality) $(x_3, (V_2 \cup \{c_2\}) + (U_3 \cup \{d_2\}))$ also embeds in $\mathbb{H}^p$, as required.

Case 3. $|T_1| = |T_2|$. T_1 is not isomorphic with T_2. For any tournament T with $|T| \leq |T_1|$ and any 1-type Q over T, Q is realized in $\mathbb{H}$.

DEFINITION 9. We may suppose that $\mathcal{T}$ consists exactly of the subtournaments of T_1 and T_2. Hence setting $\mathcal{T}_i = \mathcal{T} - \{T_i\} < \mathcal{T}$ for $i = 1, 2$, we may apply MT 3.$\mathcal{T}_i$. Let r be a ramsey 2-type for $\mathbb{H}$ over the $\mathcal{A}(\mathcal{T}_1)$-generic graph, and let $R = \{x_1, x_2, x_3\}$ be r-ramsey. Fix $a, b \in T_1$, $c \in T_2$, and let T_1^*, T_2^* be $T_1 - \{a, b\}$, $T_2 - \{c\}$ respectively. Let U_1, U_2, U_3 and V_1, V_2 be respectively three disjoint copies of T_1^*, and two disjoint copies of T_2^*. Let $A_0 = (\oplus_{i \leq 3} U_i + \oplus_{i \leq 2} V_i) \cup \{b_3, c_2\}$ with:

$$(b_3, U_3) \simeq (b, T_1^*); \quad (c_2, V_2) \simeq (c, T_2^*);$$

$$b_3 \perp (U_1 + U_2 + V_1 + V_2); \quad c_1 \perp (U_1 + U_2 + U_3 + V_1).$$

The type of b_3, c_2 will be chosen later. Let $A_i = A_0 \cup \{d_i\}$ for $i = 1, 2$ with:

$$(d_1, V_1) \simeq (c, T_2^*); \quad (d_i, U_i) \simeq (a, T_1^*); \quad (d_2, b_3, U_3) \simeq (a, b, T_1^*);$$

$$(d_1, U_2) \simeq (d_2, U_1) \simeq (b, T_1^*).$$

We shall form (R, A_1) and (R, A_2) so that after amalgamation one of the following will be isomorphic to $\mathbb{A}$:

$$(x_1, (V_1 \cup \{d_1\}) + (U_3 \cup \{b_3, d_2\}));$$

$$(x_2, (U_1 \cup \{d_1, d_2\}) + (V_2 \cup \{c_2\}));$$

$$(x_3, (U_2 \cup \{d_1, d_2\}) + (V_2 \cup \{c_2\}));$$

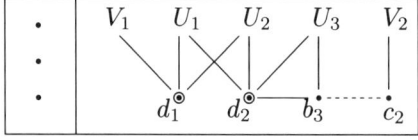

By the choice of R and the case assumption, it suffices to construct (R, A_2). This is formed by amalgamating $(R, (U_1 + U_2 \cup \{d_2\}) + V_1 + (V_2 \cup \{c_2\}))$ and (x_1, A_2). The former subfactor is afforded by the choice of R, and a suitable form of (x_1, A_2) is constructed by amalgamating $(x_1, V_1 + (U_3 \cup \{d_2, b_3\}))$ with $(\emptyset, A_2)$. Now $(\emptyset, A_2)$ will be afforded with some choice of tp$(b_3 c_2)$ by an amalgamation whose factors are provided by Theorem 7.6. The factor $(x_1, V_1 + (U_3 \cup \{d_2, b_3\}))$ is available by the minimization of $|T_2|$.

This completes the the analysis of all cases in which $T_1 \not\simeq T_2$, in a minimal counterexample to Theorem 7.8.

8.4. Theorem 7.8, conclusion

Our hypotheses remain those of the previous section. We continue our analysis of a minimal counterexample to Theorem 7.8 under the assumption that $T_1 \simeq T_2 \simeq T$, that is we have an $\mathbb{H}$-constrained 1-type $\mathbb{A}$ over $T + T$ not realized in $\mathbb{H}$, while every $\mathbb{H}$-constrained 1-type over $T + T_0$ embeds in $\mathbb{H}$ if $|T_0| < |T|$. Furthermore as stated at the outset, $\mathbb{H}$ will be taken to realize as few 1-types as possible over T.

LEMMA 8.3. $\mathcal{A}(\mathcal{T}) \implies \mathcal{A}(\mathcal{T}) + \mathcal{A}(\mathcal{T})$.

8.4. THEOREM 7.8, CONCLUSION

PROOF. Let $A_1, A_2 \in \mathcal{A}(\mathcal{T})$. We claim that $\mathcal{A}(\mathcal{T}) \implies A_1 + A_2$. If $A_1 \simeq I_1 + P_3$ then either Theorem 7.6 applies, or $A_1 + A_2$ is contained in $I_1 + P_3 + I_1 + P_3$, and Lemma 26 handles this case.

Accordingly we may take $A_i = T_i \in \mathcal{T}$, for $i = 1, 2$. Unless $T_1 \simeq T_2$ this is covered by the cases of Theorem 7.8 treated in the preceding section. So let $T_1 \simeq T_2 \simeq T$, and let H be a homogeneous directed graph embedding all elements of $\mathcal{A}(\mathcal{T})$.

Fix $a, b \in T$. Let $U; V_1, V_2; W_1, W_2$ be respectively disjoint copies of: T; $T - \{a\}$ twice; $T - \{a, b\}$ twice. Let $A_0 = (U \cup V_1 \cup V_2) + W_1 + W_2$ where the edges of A_0 are those found in its components together with two additional edges which connect the element $b_U \in U$ corresponding to b to the elements $b_i \in V_i$ $(i = 1, 2)$ corresponding to b. The orientations of these edges will be determined later. Let $A_i = A_0 \cup \{c_i\}$ for $i = 1, 2$ with:

$$(c_i, V_i) \simeq (a, T - \{a\}); \quad (c_i, W_i) \simeq (a, T - \{a, b\});$$

$$(c_1, W_2) \simeq (c_2, W_1) \simeq (b, T - \{a, b\}).$$

Amalgamation of A_1 with A_2 over A_0 will produce a copy of $T + T$. It remains to embed suitable variants of A_1, A_2 into H.

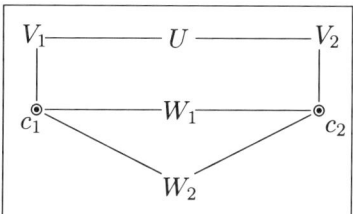

If a variant of A_1 embeds in H in which the two extra edges both originate at b_U, then it is naturally isomorphic to the corresponding form of A_2, and the same applies if both of these edges terminate at b_U. Assuming that neither of these variants embeds in H, and that $T + T$ is also omitted, then each of the following amalgamations has a unique solution:

1. $U \cup (V_1 - \{b_1\}) \cup \{c_1\} \cup W_1 \cup W_2 \cup V_2$ with $b_U \longrightarrow b_2$, amalgamated with $(U - \{b_U\}) \cup V_1 \cup \{c_1\} \cup W_1 \cup W_2 \cup V_2$;
2. $U \cup V_1 \cup W_1 \cup W_2 \cup \{c_2\} \cup (V_2 - \{b_2\})$ with $b_1 \longrightarrow b_U$, amalgamated with $(U - \{b_U\}) \cup V_1 \cup W_1 \cup W_2 \cup \{c_2\} \cup V_2$.

The result is a pair of matching factors A_1, A_2 for our intended amalgamation. Thus it suffices to check that all the factors in $(1, 2)$ embed in H, which holds by Theorem 7.6. $\square$

LEMMA 8.4. *Let P_1, P_2 be two 1-types over T realized in $\mathbb{H}$. Then there is a 1-type Q over T which is realized in $\mathbb{H}^{P_1}$ and in $\mathbb{H}^{P_2}$.*

PROOF. Let T_1, T_2, T_3 be three disjoint copies of T. Let $A_0 = T_1 \cup T_2$ with at most one extra edge joining some vertex $a \in T_1$ with some $b \in T_2$; in other words, $\text{tp}(ab)$ will be specified later. Let $\mathbb{A}_1 = (x, A_0)$ with $\text{tp}(x/T_i) = P_i$ for $i = 1, 2$, and let $\mathbb{A}_2 = (\emptyset, A_0 + T_3)$. An amalgamation of $\mathbb{A}_1$ with $\mathbb{A}_2$ will produce a suitable 1-type (x, T_3). It remains to be seen that $\mathbb{A}_1$ and $\mathbb{A}_2$ embed into $\mathbb{H}$.

$\mathbb{A}_1$ will be obtained in some form by amalgamating $(x, A_0 - \{a\})$ with $(x, A_0 - \{b\})$. It remains to check that H_2 embeds the corresponding configuration $A_0 + T_3$.

By Lemma 8.3 it suffices to check that H_2 embeds $T_1 \cup T_2$ with one additional edge, say from a to b. Let $c \in H_2$ and set $\mathbb{H}^* = (c', c^\perp)$. We shall show that $\mathbb{H}^*$ embeds $(a, A_0 - \{a\})$. As $A_0 - \{a\} = (T_1 - \{a\}) + T_2$, by the cases of Theorem 7.8.$\mathcal{T}$ already treated it suffices to prove that $\mathbb{H}^*$ embeds $(a, T_1 - \{a\})$ and (a, T_2). Reformulated in terms of H_2 these embeddings will follow from Lemma 7.6. $\square$

LEMMA 8.5. *Let P, Q be 1-types over T realized in $\mathbb{H}$. If $P + P$ embeds in $\mathbb{H}$, then $P + Q$ embeds in $\mathbb{H}$.*

PROOF. Let Q^* be a 1-type over T which is realized in both $\mathbb{H}^P$ and $\mathbb{H}^Q$. Assume that $\mathbb{H}$ omits $P + Q$. As $\mathbb{H}^P$ realizes fewer types over T than $\mathbb{H}$ does, by induction Theorem 7.8.$\mathcal{T}$ applies to $\mathbb{H}^P$. We claim that Theorem 2.$\mathcal{T}$ also applies to $\mathbb{H}^P$. In the derivation of Theorem 2.$\mathcal{T}$ given in the previous chapter, the applications of Theorem 23.$\mathcal{T}$ needed will all take place within $\mathbb{H}^P$, except within the proof of Lemma 7.15, which is contained in Lemma 8.3. Thus Theorem 2.$\mathcal{T}$ applies to $\mathbb{H}^P$, and therefore there is a ramsey type r for $\mathbb{H}^P$ over the elements of $\mathcal{A}(\mathcal{T})$. Let $R = \{x_1, x_2, x_3\}$ be r-ramsey of order 3. Any $\mathbb{H}^P$-constrained 2-directed graph of the form (R, A) with $\mathcal{A}(\mathcal{T}) \implies A$ will embed in $\mathbb{H}^P$, and a fortiori in $\mathbb{H}$.

Fix $a, b \in T$ and let $U; V_1, V_2; W_1, W_2$ be disjoint copies of $T; T - \{a\}; T - \{a, b\}$ respectively. Let $A_0 = U + V_1 + V_2 + W_1 + W_2$, $A_i = A_0 \cup \{c_i\}$ for $i = 1, 2$ with:

$$c_i \perp U; \quad (c_i, V_i) \simeq (a, T - \{a\}); c_1 \mid V_2, c_2 \perp V_1;$$

$$(c_1, c_2, W_1), (c_2, c_1, W_2) \simeq (a, b, T - \{a, b\}), \text{ apart from tp}(c_1 c_2).$$

Observe that $\mathcal{A}(\mathcal{T}) \implies A_1$ by Lemma 8.3 and Theorem 7.7.$\mathcal{T}$.

We shall amalgamate configurations of the form (R, A_1) and (R, A_2) in $\mathbb{H}$ to force a realization of $P + Q$. We require:

(1) $\text{tp}(x_1/U; V_1 \cup \{c_1\}; V_2 \cup \{c_2\}) = Q^*; P; Q$ respectively;
(2) $\text{tp}(x_i/U) = P$ for $i = 2, 3$;
(3) $\text{tp}(x_i/W_i \cup \{c_1, c_2\}) = Q$ after inserting a suitable edge joining c_1 and c_2;
(4) $\text{tp}(x_i/V_1 \cup \{c_i\})$ is realized in $\mathbb{H}^P$ for $i = 2, 3$.

Suppose that we have constructed (R, A_2) embedding in $\mathbb{H}$, and satisfying:

(5) $(R, \{c_1\} \cup (W_1 + W_2))$ is $\mathbb{H}$-constrained.

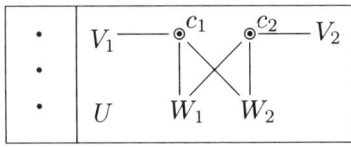

We claim that (R, A_1) is then $\mathbb{H}^P$-constrained. $(R, U + (V_1 \cup \{c_1\}))$ is $\mathbb{H}^P$-constrained by (1,2,4). (R, V_2) is contained in (R, A_2), so $(R, \{c_1\} \cup (W_1 + W_2 + V_2))$ is $\mathbb{H}$-constrained by (5), and hence also $\mathbb{H}^P$-constrained by the previous cases of Theorem 7.8. So to complete the argument it suffices to build (R, A_2) in $\mathbb{H}$ subject to (1-5).

Let $\text{tp}(x_i/V_1 \cup \{c_1\})$ be specified for $i = 2, 3$ arbitrarily subject to (4), and the constraint on $\text{tp}(x_i/c_1)$ contained in (3). We shall refer to this more specific constraint as "constraint (4)" as well. We now construct a configuration (R, A_2) in $\mathbb{H}$ compatible with clauses $(1-5)$ by amalgamating suitable factors $(R, A_2 - V_2)$ and (x_1, A_2). Here $(R, A_2 - V_2)$ is chosen so that $(R, \{c_1, c_2\} \cup (W_1 + W_2))$ is $\mathbb{H}$-constrained (taking $c_1 \perp c_2$ momentarily). This is easy to do.

8.4. THEOREM 7.8, CONCLUSION

We still have to describe (x_1, A_2). Let $P^* = Q^* + P \upharpoonright V_1$, as a type over $U + V_1$. The constraints on (x_1, A_2) are:

$$\operatorname{tp}(x_1/U + V_1) = P^*;$$

$\operatorname{tp}(x_1/\{c_1, c_2\} \cup (W_1 + W_2))$ is $\mathbb{H}$-constrained if $c_1 \perp c_2$;

$$\operatorname{tp}(x_1/c_1) = P \upharpoonright c_1; \operatorname{tp}(x_1/c_2 V_2) = Q.$$

Working in $\mathbb{H}^{P^*}$, which is also ample homogeneous embedding P and Q, we set $A = \{c_1, c_2\} \cup (W_1 + W_2 + V_2)$ and form (x_1, A) by amalgamating $(x_1, \{c_1\} + (\{c_2\} \cup V_2))$ with $(\emptyset, A)$. Both embed in $\mathbb{H}^{P^*}$, the latter by Theorem 7.7.$\mathcal{T}$. □

NOTATION 6. *For the remainder of the proof we fix a 1-type P over T such that P embeds in $\mathbb{H}$ and $P + P$ does not embed in $\mathbb{H}$.*

LEMMA 8.6. *Let $A = T \cup J$ be a finite directed graph such that for some $a \in T$, $J \perp T - \{a\}$. If A is $\mathcal{T}$-constrained and $J \cup \{a\}$ omits T, then every $\mathbb{H}$-constrained 1-type over A is realized in $\mathbb{H}$.*

PROOF. We proceed by induction on $|J|$, starting at $J = \emptyset$. We may suppose that $|T| > 2$. Let $b, c \in T - \{a\}$ and $d \in J$. Let J_0, J_1, J_2 be disjoint copies of $J - \{d\}, J, J$ and let T_0, T_1, T_2 be disjoint copies of $T - \{b\}, T - \{b,c\}, T - \{b,c\}$. Let $A_0 = (T_0 \cup J_0) \cup ((T_1 \cup J_1) + (T_2 \cup J_2))$ with $T_i \perp J_j$ for $i \neq j$. Let $c_0 \in T_0$ correspond to $c \in T$ and take $T_0 - \{c_0\} \perp T_1, T_2$. We shall determine $\operatorname{tp}(c_0/T_1 T_2)$ later. Let $A_i = A_0 \cup \{e_i\}$ for $i = 1, 2$ with:

$$(e_i, T_i \cup J_i) \simeq (b, A - \{b, c\}) \text{ for } i = 1, 2;$$

$$(e_1, T_2 \cup J_2) \simeq (e_2, T_1 \cup J_1) \simeq (c, A - \{b, c\});$$

$$(e_1, T_0 \cup J_0) \simeq (b, A - \{b, d\}); \quad (e_2, T_0 \cup J_0) \simeq (d, A - \{b, d\}).$$

Let $(x, A_1), (x, A_2)$ be taken so that after amalgamation for some $i = 0, 1,$ or 2, the configuration $(x, \{e_1, e_2\} \cup T_i \cup J_i)$ is a realization of Q. It remains to construct suitable factors $(x, A_1), (x, A_2)$ in $\mathbb{H}$.

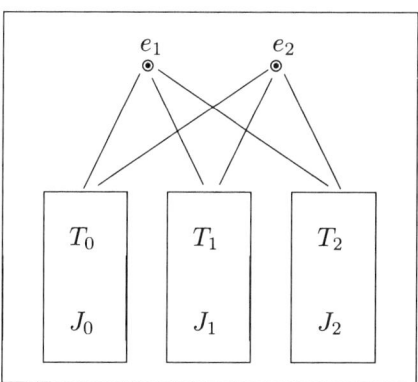

To construct (x, A_1), amalgamate $(x, A_1 - \{c_0\})$ with $(x, A_1 - (T_1 + T_2))$. Both are in any case $\mathbb{H}$-constrained, and we claim that both embed in $\mathbb{H}$. As $A_1 - \{c_0\}$ is $\mathcal{T}$-constrained and omits T, $(x, A_1 - \{c_0\})$ embeds in $\mathbb{H}$. Now $A_1 - (T_1 + T_2) = (\{e_1\} \cup T_0 \cup J_0) + J_1 + J_2$. Let $Q_0 = \operatorname{tp}(x/J_1 + J_2)$. Working in $\mathbb{H}^{Q_0}$, it suffices to embed $(x, \{e_1\} \cup T_0 \cup J_0)$. Since $|J_0| < |J|$ this can be done by induction; it suffices to check that $(x, \{e_1\} \cup T_0 \cup J_0)$ is $\mathbb{H}^{Q_0}$-constrained. For example, to see that $(x, \{e_1\} \cup T_0)$ embeds in $\mathbb{H}^{Q_0}$, let $Q_1 = Q \restriction \{e_1\} \cup T_0$ and work in $\mathbb{H}^{Q_1}$.

Now we claim that the corresponding factor (x, A_2) is $\mathbb{H}$-constrained, and that A_2 involves only tournaments of order less than $|T|$, so that (x, A_2) embeds in $\mathbb{H}$ by induction. In fact every tournament with at least three vertices contained in A_2 is contained in one of the following:

$$J_0 \cup \{a_0, e_2\}; \quad J_1 \cup \{a_1\}; \quad J_2 \cup \{a_1\}; \quad T_0;$$
$$T_1 \cup \{c_0\}; \quad T_2 \cup \{c_0\}; \quad T_1 \cup \{e_2\}; \quad T_2 \cup \{e_2\}.$$

□

LEMMA 8.7. *Let $a \in T$. There is a cross type q such that for any 1-type Q over T whose restriction to a is q, and for any 1-type P^* over $T \cup J$ with $P^* \restriction T = P$, if $J \perp T - \{a\}$, $\{a\} \cup J$ is $\mathcal{T}$-constrained omitting T, and Q and P^* are realized in $\mathbb{H}$, then $Q + P^*$ is realized in $\mathbb{H}$.*

PROOF. Suppose not. For each cross type q we may fix a counter example $Q(q), J(q), P^*(q)$. As q varies let $T(q)$ be disjoint copies of $T - \{a\}$, and let $T^* = (\oplus_q T(q)) \cup \{a\}$ with $(a, T(q)) \simeq (a, T - \{a\})$. Let $A^* = (\oplus_q J(q) \cup T)$ with $(T, J(q))$ as in $P^*(q)$. Let $A_0 = (T^* - \{a\}) + A^*$, $\mathbb{A}_1 = (\emptyset, T^* + A^*)$, $\mathbb{A}_2 = (x, A_0)$ with:

$$\text{tp}(x/T(q)) = Q(q) \restriction T(q); \quad \text{tp}(x/T \cup J(q)) = P^*(q).$$

Amalgamation of $\mathbb{A}_1$ with $\mathbb{A}_2$ produces a contradiction.

To see that $\mathbb{A}_1$ embeds in $\mathbb{H}$ it suffices to check that $\mathcal{A}(\mathcal{T}) \implies T^*$ and A^*. In both T^* and A^*, removal of one vertex produces a $\mathcal{T}$-constrained graph omitting T, so Lemma 8.6 applies.

To see that $\mathbb{A}_2$ embeds in $\mathbb{H}$ let $Q^* = \oplus_q Q(q) \restriction T(q)$, and let $\mathbb{H}^* = \mathbb{H}^{Q^*}$. It suffices to show that (x, A^*) embeds in $\mathbb{H}^*$, which follows from the previous lemma applied to $\mathbb{H}^*$.

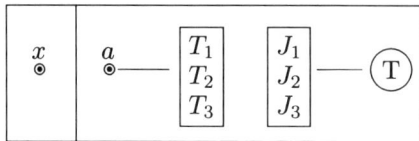

□

Now let $\mathcal{T}_0$ be the set of tournaments T_0 embedding in H_2 with $|T_0| < |T|$, and let r be a Ramsey 2-type for $\mathbb{H}^P$ over $\mathcal{A}(\mathcal{T}_0)$. Again fix $a \in T$.

LEMMA 8.8. *For any cross type q there is a 1-type Q over T whose restriction to a is q, such that for any finite configuration $\mathbb{A} = (R, A)$ with R r-ramsey, A $\mathcal{T}_0$-constrained, $\mathbb{A}$ $\mathbb{H}$-constrained, and for any $x_0 \in R$, $\mathbb{H}$ embeds a configuration of the form $(R, A+T)$ with $(R, A) \simeq \mathbb{A}$, $\text{tp}(x_0/T) = Q$, $\text{tp}(x/T) = P$ for $x \in R - \{x_0\}$.*

PROOF. Otherwise, for each type Q over T whose restriction to a is q, we have a counterexample $(R(q), A(q), x_0(q))$. Take (R, A, x_0) isomorphically embedding each of these, with $A = \oplus_q A(q)$. To get a contradiction it suffices to amalgamate $(R, A + \{a\})$ with $(R - \{x_0\}, A + T)$. The latter embeds in $\mathbb{A}$ since $(R - \{x_0\}, A)$ embeds in $\mathbb{H}^P$. □

Now we conclude the proof of Theorem 7.8.$\mathcal{T}$ by forcing $P + P$ into $\mathbb{H}$.

8.4. THEOREM 7.8, CONCLUSION

DEFINITION 10.

P is the given 1-type over T.
$a, b \in T$.
T_0 contains the tournaments T_0 embedding in H_2 with $|T_0| < |T|$.
r is ramsey for $\mathbb{H}^P$ over $\mathcal{A}(T_0)$.
$R = \{x_0, x_1, x_2\}$ is r-ramsey.
$U; V_1, V_2; W_1; W_2$ are copies of $T; T - \{a\}; T - \{a, b\}$.
$A_0 = U + V_1 + V_2 + W_1 + W_2$;
$A_i = A_0 \cup \{c_i\}$ with:

$$(c_i, V_i) \simeq (a, T - \{a\}); \quad (c_i, W_i) \simeq (a, T - \{a, b\});$$

$$c_1 \perp V_2, c_2 \perp V_1; (c_1, W_2), (c_2, W_1) \simeq (b, T - \{a, b\}).$$

q is a cross type afforded by Lemma 8.7;
Q is a 1-type over T afforded by Lemma 8.8 applied to q.

Now to force $P+P$ into $\mathbb{H}$ it suffices to amalgamate (R, A_1) with (R, A_2) where:

$$\text{tp}(x_0/U) = Q; \quad \text{tp}(x_0/V_i \cup \{c_i\}) = P \text{ for } i = 1, 2;$$

and for $i = 1, 2$:

$$\text{tp}(x_i/U) = P; \text{tp}(x_i/W_i \cup \{c_1, c_2\}) = P \text{ apart from } \text{tp}(c_1 c_2).$$

It must be checked that suitable factors (R, A_1) and (R, A_2) are available in $\mathbb{H}$. We claim that for any cross types p_1, p_2:

(*) There are types P_1, P_2 over $V_1 \cup \{c_1\}$ which are realized in $\mathbb{H}$, with $P_i \restriction c_i = p_i$, such that for any $\mathbb{H}$-constrained configuration $\mathbb{A}$ of the form $(R, U + (V_1 \cup \{c_1\} \cup J))$ with:
$$\text{tp}(x_0/U) = Q; \quad \text{tp}(x_0/V_1 \cup \{c_1\}) = P; \quad \text{tp}(x_i/U) = P;$$
$$\text{tp}(x_i/V_1 \cup \{c_1\}) = P_i (i = 1, 2); \quad J \perp (U + V_1); \quad c_1 \perp U;$$
$$\{c_1\} \cup J \text{ } T\text{-constrained omitting } T$$
we have: $\mathbb{A}$ embeds in $\mathbb{H}$.

We shall apply (*) in two forms. We first apply (*) with $p_i = \text{tp}(x_i/c_1)$ to get types P_{11}, P_{21} over $V_1 \cup \{c_1\}$. Then working over $V_2 \cup \{c_2\}$ in place of $V_1 \cup \{c_1\}$, and taking $p_i = \text{tp}(x_i/c_2)$, we get types P_{12}, P_{22} over $V_c \cup \{c_2\}$. It will then suffice to take $\text{tp}(x_i/V_j \cup \{c_j\}) = P_{ij}$ to arrive at compatible factors $(R, A_1), (R, A_2)$ embedded in $\mathbb{H}$.

To prove (*), suppose on the contrary that for every choice of P_1, P_2 we have a counterexample:

$$\mathbb{A}(P_1, P_2) = (R, U + [V_1 \cup \{c_1\} \cup J(P_1, P_2)])$$

Let $(R, U + [\{c_1\} \cup J])$ with $J = \oplus_{P_1, P_2} J(P_1, P_2)$ agree with $\mathbb{A}(P_1, P_2)$ on $(R, U + [\{c_1\} \cup J(P_1, P_2)])$. Then we arrive at a contradiction by amalgamating $(R, U + (\{c_1\} \cup J))$ with $(x_0, U + (V_1 \cup \{c_1\} \cup J))$. The first factor embeds in $\mathbb{H}$ by the choice of Q, and the second factor embeds in $\mathbb{H}$ by the choice of q.

Appendix: Examples for richer languages

This appendix contains examples of amalgamation classes defined exclusively by constraints on triangles. We deal with binary languages with few 2-types in cases where all the relations are symmetric or all relations are asymmetric. The examples were enumerated systematically using a computer and it is believed that the resulting lists are complete for the cases treated, modulo relatively trivial cases which have been excluded for clarity. We omit all cases in which there is a definable equivalence relation or the class has free amalgamation. In the asymmetric situation the case of a definable partial ordering is also rather special but we have retained these cases. In addition any example can be transformed in an inessential manner by permutation of the symbols in the language, and we discard all repetitions in this sense. After these reductions, we find that in the case of three symmetric 2-types (equality is not counted) a unique example remains, while for the case of four symmetric 2-types 27 examples are found, all functioning in a manner somewhat similar to that of the example given for three 2-types. Among the latter examples one finds three which are just the generifications of the example found for three symmetric 2-types; this means one of the three types is split generically into two types. In the asymmetric case, if two 2-types are allowed (up to symmetry) then there are 12 examples in all, one of which is finite (an oriented pentagram) and six more involve a definable partial order. The other five seem to be of a new type.

The examples given below suggest others of a more complex nature, and though this material is in a rather ill-digested form, it does seem to shed some real light on the difficulties inherent in extending the methods of this Memoir to more general languages.

We discuss the first example listed below for the language with four symmetric 2-types. Any amalgamation problem compatible with the three constraints listed can be solved using either type G or type R. Indeed to block both of these types, an amalgamation problem would have to involve the first two constraints listed, and the resulting diagram cannot be completed consistently by one of the four available 2-types (nor by an identification of vertices); the presence of the third constraint – or the pair of constraints listed in its place in the ninth example – is essential for this final point. Note that the first eight examples listed are essentially identical from this point of view, as the presence of additional constraints involving only the remaining two types A, X is irrelevant to the analysis. It follows from this observation that there are infinitely many variations on this example consisting of amalgamation classes with the same constraints on triangles (hence primitive and not free). If we pass to a language with five or more symmetric 2-types we can get $2^{\aleph_0}$ primitive non-free amalgamation classes in this way, which are in fact rather close to free amalgamation classes, but with two distinguished types rather than one. (To see this, repeat example #1 with the third constraint replaced by the family TXX with T varying over all types other than G, R, X, and then add constraints not involving G, R.)

This suggests that in general the notion of free amalgamation should be significantly generalized, along the following lines. An amalgamation class of binary structures will be said to have *semifree amalgamation* if there is a proper subset of the 2-types which is adequate for the solution of any amalgamation problem. Whereas in the case of free amalgamation it is immediately clear what is at stake – the supply of amalgamation-indecomposable structures – the notion of semifree amalgamation is relatively obscure, and in particular it is not at all clear whether the basic problem of effectivity can be handled for semifree amalgamation classes. At this point one can only suggest that if the results of this Memoir apply in some form to more complex languages, some such notion is likely to be essential. One may observe that in the case of four symmetric 2-types, the classes defined by constraints on triangles (neglecting the imprimitive case) are semifree with respect to a set of just two types. Furthermore these two types do not occur together in any constraint. These examples are in fact rather transparent as a result, and further exploration will no doubt turn up rather more elaborate phenomena.

It would be of considerable interest to have a complete classification of the exceptional homogeneous structures for language with three symmetric 2-types, which means in the first place: all those with some triangle forbidden (and, inductively, all those for which the locus of some 1-type over a single point is itself exceptional). In particular in connection with example #1 one of the more striking cases would be the case of amalgamation classes satisfying the first two of the three constraints given there. One may also conjecture that there is no other imprimitive example involving triangle constraints which themselves involve all three of the available 2-types. It should be noted that the lists given below have not been proved correct, even for the case of three symmetric 2-types, and it would be good to have a reasonably efficient proof of this.

Amalgamation classes determined by constraints on triangles

In the charts below, each triple represents the sequence of 2-types in a forbidden triangle, and each set of forbidden triangles is the set of negative constraints for an amalgamation class.

Language: $\{G, R, B\}$.   1 example.

#1   *RBB   GGB   BBB*

Language: $\{G, R, A, X\}$.   27 examples.

#1	*RXX GAX AXX*
#2	*RXX GAX AXX XXX*
#3	*RXX GAX AXX AAX*
#4	*RXX GAX AXX AAA*
#5	*RXX GAX AXX AAX XXX*
#6	*RXX GAX AXX XXX AAA*
#7	*RXX GAX AAX AXX AAA*
#8	*RXX GAX AAX AXX XXX AAA*
#9	*RXX GAX AAX XXX*
#10	*RXX GAX AAX XXX AAA*
#11	*RXX GGX AXX XXX*
#12	*RXX GGX AAX AXX XXX*
#13	*RXX GGX AXX XXX AAA*
#14	*RXX GGX AAX AXX XXX AAA*
#15	*RXX GAX GGX AXX XXX*
#16	*RXX GAX GGX AAX AXX XXX*
#17	*RXX GAX GGX AXX XXX AAA*
#18	*RXX GAX GGX AAX AXX XXX AAA*
#19	*RXX GAX GGX AAX XXX*
#20	*RXX GAX GGX AAX XXX AAA*
#21	*RAA RXX GAX AAX XXX*
#22	*RAA RXX GAX AAX AXX*
#23	*RAA RXX GAX AAX AXX XXX*
#24	*RAA RXX GAX AXX XXX AAA*
#25	*RAA RXX GAX AAX AXX XXX AAA*
#26	*RRX RAA RXX GAX GXX AAX XXX*
#27	*RRA RRX GAA GAX GXX AAX AXX XXX AAA*

APPENDIX: EXAMPLES FOR RICHER LANGUAGES

Language: $\{R^\pm, G^\pm\}$ (2 asymmetric 2-types): 12 examples.

Finite (order 5):

#1 $R^+R^+R^+$ $R^+R^-R^+$ $G^+G^+G^+$ $G^+G^-G^+$ $R^+G^+G^+$ $R^+G^-G^+$ $R^+G^-G^-$ $G^+R^+R^+$
   $G^+R^+R^-$ $G^+R^-R^-$

Transitive:

#2 $G^+G^-G^+$ $R^+G^+G^-$ $R^+G^-G^+$
#3 $G^+G^-G^+$ $R^+G^+G^-$ $R^+G^-G^+$ $G^+R^-R^+$
#4 $R^+R^-R^+$ $G^+G^-G^+$ $R^+G^-G^+$ $G^+R^-R^+$
#5 $R^+R^-R^+$ $G^+G^-G^+$ $R^+G^-G^+$ $G^+R^+R^-$ $G^+R^-R^+$
#6 $R^+R^-R^+$ $G^+G^+G^+$ $G^+G^-G^+$ $R^+G^-G^+$ $G^+R^-R^+$
#7 $R^+R^-R^+$ $G^+G^-G^+$ $R^+G^+G^-$ $R^+G^-G^+$ $G^+R^+R^-$ $G^+R^-R^+$

General type:

#8 $G^+G^-G^+$ $R^+G^-G^+$
#9 $G^+G^+G^+$ $R^+G^-G^+$
#10 $G^+G^+G^+$ $G^+G^-G^+$ $R^+G^-G^+$
#11 $R^+R^-R^+$ $G^+G^+G^+$ $G^+G^-G^+$ $G^+R^+R^+$ $G^+R^-R^-$
#12 $R^+R^-R^+$ $G^+G^+G^+$ $R^+G^-G^+$ $G^+R^+R^+$ $G^+R^+R^-$ $G^+R^-R^-$ #

Bibliography

[1] Reinhold Baer. Die Kompositionsreihe der Gruppe aller eineindeutigen Abbildungen einer unendlichen Menge auf sich. *Studia Math.*, **5**:15-17, 1935.

[2] James Bennett. Reducts of some binary homogeneous structures. Doctoral thesis, Rutgers University, 1993.

[3] Chantal Berline. Elimination of quantifiers for non semi-simple rings of characteristic p. In *Model Theory of Algebra and Arithmetic*, volume 834, pages 10-19, New York, 1980. Springer. MR 82k:03034.

[4] Chantal Berline and Gregory Cherlin. QE rings in characteristic p. In *Logic Year 1979/80*, volume 859, pages 16-31. Springer, 1981. MR 84a:03032.

[5] Chantal Berline and Gregory Cherlin. QE rings in characteristic p^n. *J. Symb. Logic*, **48**:140-162, 1983. MR 84f:03023.

[6] Maurice Boffa, Angus Macintyre, and Françoise Point. The quantifier elimination problem for rings without nilpotent elements and for semisimple rings. In *Model Theory of Algebra and Arithmetic*, volume 834, pages 20-30, Berlin, 1980. Springer. MR 82d:03054.

[7] Peter J. Cameron. On a theorem of Livingstone and Wagner. *Math. Z.*, **137**:343-350, 1974. MR 50 #4712.

[8] Peter J. Cameron. Cohomological aspects of two-graphs. *Math. Z.*, **157**:101-119, 1977. MR 81a:05061.

[9] Peter J. Cameron. Orbits of permutation groups on unordered sets I. *J. London Math. Soc.*, **17**:410-414, 1978. MR 58 #11136.

[10] Peter J. Cameron. Orbits of permutation groups on unordered sets II. *J. London Math. Soc.*, **23**:249-264, 1981. MR 82k:20005.

[11] Peter J. Cameron. Orbits of permutation groups on unordered sets IV. Homogeneity and transitivity. *J. London Math. Soc.*, **23**:249-264, 1981. MR 84f:20007b.

[12] Peter J. Cameron. Orbits of permutation groups on unordered sets III. Imprimitive groups. *J. London Math. Soc.*, **27**:238-247, 1983. MR 84f:20007a.

[13] Peter J. Cameron. Some treelike objects. *Quarterly J. of Math., Series 2*, **38**:155-183, 1987. MR 89a:05009.

[14] Peter J. Cameron. *Oligomorphic Permutation Groups*. number 152 in London Mathematical Society Lecture Notes. Cambridge University Press, Cambridge, UK. MR 92f:20002.

[15] Peter J. Cameron and Simon Thomas. Groups acting on unordered sets. *Proc. Amer. Math. Soc.*, **59**:541-557, 1989. MR 91g:20003.

[16] Gregory Cherlin. Aspects of $\aleph_0$-categoricity. In *Proceedings of the International Congress on Logic, Philosophy, Methodology, and History of Science, VII (Salzburg)*, pages 99-114, Amsterdam, 1986. North-Holland. Studies in Logic and the Foundations of Mathematics, 114. MR 87m:03046.

[17] Gregory Cherlin. Homogeneous directed graphs I. The imprimitive case. In Paris Logic Group, editor, *Logic Colloquium 1985*, pages 67-88, New York, 1987. North-Holland. MR 88d:03074.

[18] Gregory Cherlin. Homogeneous tournaments revisited. *Geometria Dedicatae*, **26**:231-240, 1988. MR 89k:05039.

[19] Gregory Cherlin. Combinatorial problems connected with finitely homogeneous structures. In *Proceedings of the International Conference on Algebra, Dedicated to the memory of A. I. Mal'tsev*, L. Bokut, Yu. Ershov, and A. Kostrikin, eds., pp. 3-30, Contemporary Mathematics **131**, AMS, Providence, RI, 1992. MR 93m:03054.

[20] Gregory Cherlin and Ulrich Felgner. Homogeneous finite groups. In preparation.

[21] Gregory Cherlin and Ulrich Felgner. Homogeneous solvable groups. *J. London Mathematical Society*, **44**:102-120, 1991. MR 92m:20027.

[22] Gregory Cherlin and Alistair H. Lachlan. Finitely homogeneous relational structures. *Trans. Amer. Math. Soc.*, **296**:815-850, 1986. MR 88f:03023.

[23] Cheryl Chute Miller. Imprimitive automorphism groups. *Quart. J. Math. Oxford Ser. (2)*, **43**:23-44, 1992. MR 93a:20005.

[24] John Dixon, Peter Neumann, and Simon Thomas. Subgroups of small index in infinite symmetric groups. *Bull. London Math. Soc.*, **18**:580-586, 1986. MR 88i:20004.

[25] Manfred Droste. *Structure of Partially Ordered Sets with Transitive Automorphism Groups.* Vol. 334 of *Memoirs of the American Mathematical Society*, AMS, Providence, RI, 1985. MR 87d:06005

[26] Manfred Droste and H. Dugald Macpherson. On k-homogeneous posets and graphs *J. Combinatorial Theory, Series A*, **56**:1-15. MR 92e:03049.

[27] Manfred Droste and John Truss. Subgroups of small index in ordered permutation groups. *Quarterly J. of Math., Series 2*, **42**:31-47, 1991. MR 92g:06025.

[28] Paul Erdős, Daniel Kleitman, and Bruce Rothschild. Asymptotic enumeration of K_n-free graphs. In *Colloquio Internazionale sulle Teorie Combinatorie (Rome, 1973)*, Tomo II, pages 19-27, 1976. MR 57 #2984.

[29] David Evans. Subgroups of small index in infinite general linear groups. *Bull. London Math. Soc.*, **18**:587-590, 1986. MR 88i:20005.

[30] Ronald Fagin. Probabilities on finite models. *J. Symb. Logic*, **41**:50-58, 1976. MR 57 #16042.

[31] Roland Fraïssé. Sur certaines relations qui généralisent l'ordre des nombres rationnels. *C. R. Acad. Sci. Paris*, **237**:540-542, 1953. MR 15 #192.

[32] Roland Fraïssé. Sur l'extension aux relations de quelques propriétés connues des ordres. *C. R. Acad. Sci. Paris*, **237**:508-510, 1953. MR 57 # 16042.

[33] Roland Fraïssé. Sur l'extension aux relations de quelques propriétés des ordres. *Ann. Ecole Norm. Sup.*, **71**:361-388, 1954. MR 16 #1006.

[34] Roland Fraïssé. Sur quelques classifications des systèmes de relations. *Publications Scientifiques, Université d'Alger, Série A*, **1**:35-182, 1954.

[35] Roland Fraïssé. *Théorie des relations*, volume 118 of *Studies in Logic and the Foundations of Mathematics*. North-Holland, New York, 1986. MR 87f:03139.

[36] Anthony Gardiner. Homogeneous graphs. *J. Combin. Theory Ser. B*, **20**:94-102, 1976. MR 52 #7316.

[37] Edward Gaughan. The index problem for infinite symmetric groups. *Proc. Amer. Math. Soc.*, **15**:527-528, 1964. MR 29 #4789.

[38] Yu. Glebski, D. Kogan, M. Liogon'kiĩ, , and V. Talanov. Volume and fraction of satisfiability of formulas of the lower predicate calculus. *Kibernetika*, **2**:17-27, 1969. MR 46 #42.

[39] Ya. Golfand and Mikhail Klin. On k-homogeneous graphs. In *Algorithmic Studies in Combinatorics*, pages 76-85, Moscow, 1978. Nauka. MR 80d:05043.

[40] Johannes Guttwerden. Die Hruschowskigeometrien. in *Proceedings of the 7th Easter Conference on Model Theory*, Berndt Dahn and Helmut Wolter, eds., 106-118, 1989.

[41] Ronald Graham, Bruce Rothschild, and Joel Spencer. Ramsey Theory. Second edition. Wiley-Interscience series in Discrete Mathematics and Optimization, John Wiley, New York, 1990. xii+196 pp. MR 90m:05003.

[42] C. Ward Henson. A family of countable homogeneous graphs. *Pacific J. Math.*, **38**:69-83, 1971. MR 46 #3377.

[43] Wilfrid Hodges. *Building Models by Games*. Number 2 in *London Mathematical Society Student Texts*, Cambridge University Press, Cambridge, UK, 1985. MR 97h:03045.

[44] Ehud Hrushovski. A new strongly minimal set. *Annals of Pure and Applied Logic*, **62**:147-166, 1993. MR 94d:03064.

[45] Ehud Hrushovski. Finite structures with few types. Vol. 411 of *NATO ASI Series C: Mathematical and Physical Sciences*, Kluwer, Dordrecht, 175-187. Proceedings, Banff, 1991. MR 95h:03084.

[46] Bjarni Jónsson. Universal relational systems. *Mathematica Scandinavica*, **4**:193-208, 1956. MR 20 #3091.

[47] Bjarni Jónsson. Homogeneous universal relational systems. *Mathematica Scandinavica*, **8**:137-142, 1960. MR 49 #10625.

[48] William Kantor. On incidence matrices of finite projective and affine spaces. *Math. Z.*, **124**:315-318, 1972. MR 59 #13850.

[49] William Kantor, Martin Liebeck, and Dugald Macpherson. $\aleph_0$-categorical structures smoothly approximated by finite substructures. *Proc. London Math. Soc.*, **59**:439-463, 1989. MR 91e:03033.
[50] Richard Kaye and H. Dugald Macpherson. *Automorphisms of first-order structures*. Oxford University Press, Oxford, UK, 1994. MR 96d:03048
[51] Julia Knight and Alistair H. Lachlan. Shrinking, stretching, and codes for homogeneous structures. In John Baldwin, editor, *Classification Theory (Chicago, 1985)*, volume 1292 of *Lecture Notes Math.*, 1987. MR 90k:03033.
[52] Phokion Kolaitis, H. Prömel, and Bruce Rothschild. Asymptotic enumeration and a 0-1 law for m-clique free graphs. *Bull. Amer. Math. Soc.*, 13:160-162, 1985. MR 87c:05068.
[53] Phokion Kolaitis, H. Prömel, and Bruce Rothschild. K_{l+1}-free graphs: Asymptotic structure and a 0-1 law. *Trans. Amer. Math. Soc.*, **303**:637-671, 1987. MR 88j:05016.
[54] David Kueker and M. Chris Laskowski. On generic structures. *Notre Dame Journal of Formal Logic*, *33*:175-183. MR 93k:03032.
[55] Alistair H. Lachlan *Obituary*: Alan Mekler. *Order* 89:99-101, 1992. MR 93m:01074.
[56] Alistair H. Lachlan. Finite homogeneous simple digraphs. In Jacques Stern, editor, *Logic Colloquium 1981*, volume 107 of *Studies in Logic and the Foundations of Mathematics*, pages 189-208, New York, 1982. North-Holland. MR 85h:05049.
[57] Alistair H. Lachlan. Countable homogeneous tournaments. *Trans. Amer. Math. Soc.*, **284**:431-461, 1984. MR 85i:05118.
[58] Alistair H. Lachlan. On countable stable structures which are homogeneous for a finite relational language. *Israel J. Math.*, **49**:69-153, 1984. MR 87h:03047a.
[59] Alistair H. Lachlan. Homogeneous structures. In Andrew Gleason, editor, *Proceedings of the ICM 1986*, pages 314-321, Providence, RI, 1987. AMS. MR 89d:03030.
[60] Alistair H. Lachlan and Saharon Shelah. Stable structures homogeneous for a finite binary language. *Israel J. Math.*, **49**:150-180, 1984. MR 87h:03047b.
[61] Alistair H. Lachlan and Robert Woodrow. Countable ultrahomogeneous undirected graphs. *Trans. Amer. Math. Soc.*, **262**:51-94, 1980. MR 82c:05083.
[62] Daniel Lascar. Autour de la propriété du petit indice. Proceedings of the London Mathematical Society, Series 3, **62**:25-53. MR 92f:03029.
[63] Daniel Lascar. The group of automorphisms of a relational saturated structure Vol. 411 of *NATO ASI Series C: Mathematical and Physical Sciences*, Kluwer, Dordrecht, 225-236, Proceedings, Banff, 1991. MR 95d:03060
[64] Daniel Lascar and Saharon Shelah. Uncountable saturated structures have the small index property. *Bulletin of the London Mathematical Society*, **25**:125-131, 1993. MR 94d:03068.
[65] Donald Livingstone and Ascher Wagner. Transitivity of permutation groups on unordered sets. *Math. Z.*, **90**:393-403, 1967. MR 32 #4183.
[66] Angus Macintyre, Kenneth McKenna, and Laurentius Petrus Dignus van den Dries. Elimination of quantifiers in algebraic structures. *Adv. Math.*, **47**:74-87, 1983. MR 84f:03028.
[67] H. Dugald Macpherson. Groups of automorphisms of $\aleph_0$-categorical structures. *Quarterly J. of Math., Series 2*, **37**:449-465. MR 88d:20007
[68] S. McLeish. *The sufficiency of going forth in first-order homogeneous structures*. Doctoral thesis, Queen Mary and Westfield College, 1994.
[69] Peter Neumann. Personal communication.
[70] Jaroslav Nešetřil and Vojtěch Rödl. The Ramsey property for graphs with forbidden complete subgraphs. *J. Combin. Theory Ser. B*, **20**:243-249, 1976. MR 54 #133.
[71] Jaroslav Nešetřil and Vojtěch Rödl. Partitions of finite relational and set systems. *J. Combinatorial Theory, Series A*, **22**:289-312, 1977. MR #55 10283.
[72] Jaroslav Nešetřil and Vojtěch Rödl. Ramsey classes of set systems. *J. Combinatorial Theory, Series A*, **34**:183-201, 1983. MR 84i:05016.
[73] Maurice Pouzet. *Sur la théorie des relations*. Doctoral thesis, Lyon, 1978.
[74] Alex Rosenberg. The structure of the infinite general linear group. *Ann. Math.*, **68**:278-294, 1958. MR 21 #1319.
[75] Matatyahu Rubin. On the reconstruction of $\aleph_0$-categorical structures from their automorphism groups. *Proceedings of the London Mathematical Society*, **69**:225-249, 1994. MR 95e:03102.
[76] Daniel Saracino. Amalgamation bases for nil-2 groups *Algebra Universalis*, **16**:47-62. MR 84i:20035.

[77] Daniel Saracino and Carol Wood. QE nil-2 groups of exponent 4. *J. Alg.*, **76**:337-382, 1982. MR 83i:03052.

[78] Daniel Saracino and Carol Wood. QE commutative nil rings. *J. Symb. Logic*, **49**:644-651, 1984. MR 85i:03093.

[79] Daniel Saracino and Carol Wood. *Finite Quantifier-Eliminable Rings in Characteristic* 4, volume 106 of *Lect. Notes Pure Applied Math.*, pages 329-348. Dekker, New York, 1987. MR 89g:03044.

[80] Daniel Saracino and Carol Wood. Homogeneous finite rings in characteristic 2^n. *Annals Pure Applied Logic*, **40**:11-28, 1988. MR 89i:03070.

[81] Daniel Saracino and Carol Wood. Finite homogeneous rings of odd characteristic. In *Logic Colloquium '84 (Manchester, 1984)*, Studies in Logic and Foundations of Mathematics, No. 120, pages 207-224, North Holland, Amsterdam, 1986. MR 88a:03077.

[82] James Schmerl. Countable homogeneous partially ordered sets. *Alg. Univ.*, **9**:317-321, 1979. MR 81g:06001.

[83] J. Sheehan. Smoothly embeddable subgraphs. *J. London Math. Soc.*, **9**:212-218, 1974. MR 51 #229.

[84] Saharon Shelah and Simon Thomas. Implausible subgroups of infinite symmetric groups. *Bull. London Math. Soc.*, **20**:313-318, 1988. MR 90a:20009.

[85] L. Svenonius. $\aleph_0$-categoricity in first-order predicate calculus. *Theoria* **25**:82-94, 1959. MR 25 #1986a.

[86] Simon Thomas. Groups acting on infinite dimensional projective space. *J. London Math. Soc.*, **34**:265-273, 1986. MR 87k:20014.

[87] Simon Thomas. Reducts of the random graph. *J. Symbolic Logic*, **56**:176-181, 1991. MR 92m:05092.

[88] Simon Thomas. Reducts of random hypergraphs. *Annals Pure Applied Logic*, **80**:165-193, 1996.

[89] John Truss. The group of the countable universal graph. *Math. Proc. Cambridge Phil. Soc.*, 98:213-245, 1985. MR 86j:05075.

[90] John Truss. Infinite permutation groups I. Products of conjugacy classes. *J. Algebra*, **120**:454-493, 1989. MR 90c:20004.

[91] John Truss. Infinite permutation groups II. Subgroups of small index. *J. Algebra*, **120**:494-515, 1989. MR 90c:20005.

[92] H. Wielandt. *Unendliche Permutationsgruppen*. Unpublished typescript, 1959.

Index of Notation

[ ]: A[B] composition, 53
[,]: 21, 53, 76
$\Longrightarrow$, 8
$\perp$, $-$, 2-types, 53
$+$: H^+, 55, 122; $A+B$, 76
$-$: H^-, 122
^{a} : S^a, 76
$\mathcal{A}$, class of finite structures, 6
$\mathcal{A}^*$, $\mathcal{A}^+$, 81
$\mathcal{A}(n)$ 54
$\mathcal{A}_n$, set of finite directed graphs 78
$\mathcal{A}^r$, 57
$\mathcal{A}(\mathcal{T})$, 7, 119
$\mathcal{A}(T_1, T_2; \mathcal{P})$, 19
C_3, 17
C_5, 4
Δ_n, diagonal n-tournament, 18
G^c, graph complement, 53
Γ_∞, 4
$\Gamma(\mathcal{P})$, $\Gamma_k(T_1, T_2)$, 19
Γ_n, 4, 74
I_n, co-clique, 53, 74
K_n or $K(n)$, complete graph, 3, 53
$L(n)$, linear tournament, 21
$n * I_\infty$, 74
$n \cdot T = I_n[T]$, 75
P_3, path, 53; 1-type over, 54
$\mathcal{P} = (P^\circ, P^+, P^-)$, partition of cross types, 19
$\mathcal{P}$, generic partial order, 74, 75
$\mathcal{P}(3)$, generic local partial order, 74, 76
$\mathbb{Q}$, rational order, 17
Q_4, 131
$S(2)$, $S(3)$, 17, 74
$\mathbb{S}$, finite n-tournament, 25
τ, polarization type, 47
$\mathbb{T}$, n-tournament, 12
$\mathbb{T}(p)$, derived n tournament, 28
$\hat{T}$, covering of a tournament, 74
T^∞, 17, 74

Index

$\aleph_0$-categoricity 3
Alcuin 13
amalgamation,
  class 1, 2, 6-21, 24, 25, 31, 36, 48, 49, 51,
    52, 54, 57-60, 66, 75, 78, 79, 81-83, 90,
    91, 99, 100, 119, 123ff., 150-153.
    finitely constrained, 13, 15, 20, 149-152
  free, 7, 8, 11-14, 16, 75, 150
  property, 3, 9, 19, 31
  semifree, 151
  strong, 10, 11, 18
ample, 34, 56-58, 62, 63, 67, 71, 120, 122,
  124, 125, 132, 133, 136, 139, 140, 146
antichain, 7, 8, 14, 15
automation, 15
back-and-forth, 7
Baer, Reinhold, 5
Bennet, James, 5
Berline, Chantal, 3
Cameron, Peter, 6
cardinal, $2^{\aleph_0}$, v, 1, 3, 5, 7, 14.
catalog, 13
  homogeneous directed graphs, 73
  homogeneous graphs, 4
  homogeneous 2-tournaments, 17, 18
coloring
  edge, 5, 11, 14
  vertex, 1, 2, 17, 50
completion, 22, 40, 50, 88, 94, 97
composition, 21, 23, 28, 53, 74, 75
consequence, 54, 60, 61, 73, 79, 121, 130,
  132, 139
constraint, 13, 14, 15, 28, 150
  $\mathbb{H}$-constrained, 54-73, 120-136, 141-148
  see under partition
  $\mathcal{T}$-constrained, 119-124, 137-140
cut, 22, 40, 50, 88
data, canonical, 38, 45-52
Dedekind, Richard, 22, 40, 50, 88, 94, 97
degree, 120
directed graph
  composite, 74
  embedding I_∞, v, 119
  finite, 15, 16, 27, 32, 81, 100
  free, 74, 119, 138
  $\mathcal{T}$-generic, *see* directed graph, free

  homogeneous, v, 1, 2, 3, 4, 5, 6, 12, 13,
    18, 123
    catalog, 74
    exceptional, 74
    imprimitive, 2, 13, 74, 75
  omitting I_∞, v, 78, 118
  primitive, 13, 78, 86, 88, 89, 119
  wreathed, *see* directed graph, composite
  cf. tournament
2-directed graph, 12, 76, 78, 80, 94, 96, 99
  ff.
n-directed graph, 14, 75, 76, 77, 119, 121
Dixon, John, 5
domination, 17, 19, 25, 85, 96, 104
van den Dries, Laurentius Petrus Dignus, 3
effectivity, 1, 8, 9, 12-15, 20, 160
equality, 6, 9, 10, 11, 22, 85, 150
Erdős, Pàl, 16
Evans, David, 5
Fagin, Ronald, 16
Felgner, Ulrich, 3
Fraïssé, Roland, 1, 2, 3, 4, 6, 7, 8, 14, 49, 75,
  79, 83
fraternity, 18
Gardiner, Anthony, 4, 53
Gaughan, Edward, 5
generation, 8, 49, 75, 77
genericity, 6, 17, 18, 21, 22, 50, 75, 69, 150
  for $\mathcal{P}$, 19, 20, 34, 40, 47, 119, 120, 141,
    143
  $\mathcal{T}$-generic, 73, 119, 120, 141, 143
  see also:
  partial order
generification, 9-11, 18, 19, 75, 150
Glebski, Yu., 16
Golfand, Ya., 4, 53
Goode, Jonathan Barnham,
  see Pope, Alexander
graph
  automorphisms, 5, 6
  finite, 4, 15, 53, 56, 58, 67
  generic, 4, 5, 10, 11
  homogeneous, 2, 4, 53
  partitioned, *see* n-graph
  Rado, *see under* generic
  random, *see* generic

see also directed graph
n-graph, 14
  2-graph 54ff.
Grégoire, 152
group
  automorphism, 1, 2, 3, 5, 6, 74
  classical, 5
  homogeneous, 4
  nilpotent, v, 3
Henson, C. Ward, 1, 8
Higman, Donald, 15
homogeneity, v, 1-13, 17, 38, 53, 74, 119
Hrushovski, Ehud, 6
indecomposable, 7, 14, 74, 151
joint embedding, 6
Jung, C. F., 3
Kantor, William, 4, 6
Kleitman, Daniel, 16
Klin, M., 4, 53
Knight, Julia, 4
Kobaltana, see Zembla
Kogan, D., 16
Kolaitis, Phokion, 16
Kruskal, D, 15
Lachlan, Alistair H., v, 1, 2, 4, 9, 11, 13, 17,
    23, 32, 53, 63, 76, 79, 133
  see also Ramsey, Lachlan's Ramsey argument
language, 9, 10,
  binary, 11, 14, 15, 29, 150
  canonical, 18
  natural, 3
  relational, 4, 5, 9, 13
Latka, Brenda, 15
liberty, 7, 8, 11, 12-14, 26, 57, 61, 62, 75, 78,
    123, 132, 151
Liogonkiĭ, 16
local order, see under tournament
Macintyre, Angus, 3
Macpherson, Dugald, 4
McKenna, Kenneth, 3
MT (Main Theorem)
  MT1, 77, 80-81
  MT2, 77, 82-84, 99-100
Nešetřil, Jaroslav, 5
Nešetril-Rödl, 5
Neumann, Peter, 3
omit, 4, 16, 53, 57, 74
parallelogram law, 27
partial order
  dense subset, 75
  generic, 75, 85, 125, 127
partition $\mathcal{P}$, 19, 28
Poizat, Bruno
  see Goode, Jonathan Barnham
Pope, Alexander, see Zembla
primitive, 5, 10, 18, 150
problems

automation, 14, 15
classification, 3, 8, 9
effectivity, 1, 8, 9
open, 1, 13ff
probabilistic, 16
reconstruction, 5
small index, 5
Prömel, H., 16
pseudoplane, 6
quantifier elimination, 3
Rado, Richard, see graph, generic
Ramsey, Franklin P.
  Ramsey's theorem, 5, 9, 11, 23, 55, 56
  Lachlan's Ramsey argument, 35, 36, 49,
    58, 62, 78, 81, 83, 84, 94, 100, 123, 132
  type see under type, ramsey
rank, 120
realize, see under 1-type
$\mathbb{T}$-restricted,
  see under 2-tournament, n-tournament
restriction,
  by parity, 75
  of n-tournament, 12, 17, 38, 55, 120
ring, homogeneous, v, 3
Rödl, Vojtěch, 5
Rosenberg, Alex, 5
Rothschild, Bruce, 16
Rubin, Matityahu, 5
Ruckfahrt, 15
Saracino, Daniel, 4
Schmerl, James 75, 85, 125
semigenericity, 74
Shelah, Saharon, 4, 5
sink, source, see under type
small index property, 5-6
stability, 4
strongly minimal set, 6
structure
  generic, 7
  homogeneous, v, 1
  $\mathcal{L}$-structure, 10
  primitive, 2
  ramsey, 11
  random, 16
Sub(Γ), 6
Talanov, V., 16
Thomas, Simon, 5, 6
Titanic, 2
tournament,
  composite, 21
  generic, 5, 10, 17
  homogeneous, 4, 17, 74, 79, 81
  linear, 21, 76
  local order, 32, 33, 34
  random, see generic
  wreath product, see composite
2-tournament,
  constrained, 28

derived, 34, 96
  $\mathcal{P}$-generic, 19, 26, 32
  homogeneous, 17, 20
  $\mathbb{T}$-restricted, 21
  unconstrained, 33
3-tournament
  critical, 38
  derived, 39, 84
  hypercritical, 39
  small, 39
n-tournament, 6, 23
  diagonal, 18, 22
  homogeneous, 1, 2, 12, 14, 20, 38
  Lachlan's sense, 17
  polarized, 47
  $\mathbb{T}$-restricted, 38, 48
  shuffled, 18, 123
  cf. 2-tournament, 3-tournament
transversal, 12
Truss, John, 5
type, 23
  1-type, 11
    special case, 12, 57, 62, 68, 78, 80
  2-type, 9, 11, 33, 58, 74, 75, 76
  cross type, 17, 19, 55
  neutral, 26, 32
  ramsey, 62-66, 69, 70, 72-73, 133, 136, 141,
    143, 145, 147, 148
  realize, 12, 25, 76, 80
  sink, 26ff., 32
  source, 26ff.
Ulam, Stanislas, 5
va-et-vient, 15
variant, 18, 19, 26, 28, 41, 50, 76
variation of parameters, 29, 39, 40, 47, 99,
    102, 104, 107, 109, 118
Wood, Carol, v, 3, 4
Woodrow, Robert, v, 2, 4, 8, 24, 53, 56, 81,
    121
wreath product, *see* composition
  see also under directed graph, tournament
Zembla, a distant northern land,
  not in the text
Zilber, Boris, 6

Editorial Information

To be published in the *Memoirs*, a paper must be correct, new, nontrivial, and significant. Further, it must be well written and of interest to a substantial number of mathematicians. Piecemeal results, such as an inconclusive step toward an unproved major theorem or a minor variation on a known result, are in general not acceptable for publication. *Transactions* Editors shall solicit and encourage publication of worthy papers. Papers appearing in *Memoirs* are generally longer than those appearing in *Transactions* with which it shares an editorial committee.

As of September 30, 1997, the backlog for this journal was approximately 8 volumes. This estimate is the result of dividing the number of manuscripts for this journal in the Providence office that have not yet gone to the printer on the above date by the average number of monographs per volume over the previous twelve months, reduced by the number of issues published in four months (the time necessary for preparing an issue for the printer). (There are 6 volumes per year, each containing at least 4 numbers.)

A Copyright Transfer Agreement is required before a paper will be published in this journal. By submitting a paper to this journal, authors certify that the manuscript has not been submitted to nor is it under consideration for publication by another journal, conference proceedings, or similar publication.

Information for Authors and Editors

Memoirs are printed by photo-offset from camera copy fully prepared by the author. This means that the finished book will look exactly like the copy submitted.

The paper must contain a *descriptive title* and an *abstract* that summarizes the article in language suitable for workers in the general field (algebra, analysis, etc.). The *descriptive title* should be short, but informative; useless or vague phrases such as "some remarks about" or "concerning" should be avoided. The *abstract* should be at least one complete sentence, and at most 300 words. Included with the footnotes to the paper, there should be the 1991 *Mathematics Subject Classification* representing the primary and secondary subjects of the article. This may be followed by a list of *key words and phrases* describing the subject matter of the article and taken from it. A list of the numbers may be found in the annual index of *Mathematical Reviews*, published with the December issue starting in 1990, as well as from the electronic service e-MATH [**telnet e-MATH.ams.org** (or **telnet 130.44.1.100**). Login and password are **e-math**]. For journal abbreviations used in bibliographies, see the list of serials in the latest *Mathematical Reviews* annual index. When the manuscript is submitted, authors should supply the editor with electronic addresses if available. These will be printed after the postal address at the end of each article.

Electronically prepared papers. The AMS encourages submission of electronically prepared papers in $\mathcal{A}_{\mathcal{M}}\mathcal{S}$-TeX or $\mathcal{A}_{\mathcal{M}}\mathcal{S}$-LaTeX. The Society has prepared author packages for each AMS publication. Author packages include instructions for preparing electronic papers, the *AMS Author Handbook*, samples, and a style file that generates the particular design specifications of that publication series for both $\mathcal{A}_{\mathcal{M}}\mathcal{S}$-TeX and $\mathcal{A}_{\mathcal{M}}\mathcal{S}$-LaTeX.

Authors with FTP access may retrieve an author package from the Society's Internet node **e-MATH.ams.org** (130.44.1.100). For those without FTP access, the author package can be obtained free of charge by sending e-mail to pub@ams.org (Internet) or from the Publication Division, American Mathematical Society, P.O. Box 6248, Providence, RI 02940-6248. When requesting an author package, please specify $\mathcal{AMS}$-TeX or $\mathcal{AMS}$-LaTeX, Macintosh or IBM (3.5) format, and the publication in which your paper will appear. Please be sure to include your complete mailing address.

Submission of electronic files. At the time of submission, the source file(s) should be sent to the Providence office (this includes any TeX source file, any graphics files, and the DVI or PostScript file).

Before sending the source file, be sure you have proofread your paper carefully. The files you send must be the EXACT files used to generate the proof copy that was accepted for publication. For all publications, authors are required to send a printed copy of their paper, which exactly matches the copy approved for publication, along with any graphics that will appear in the paper.

TeX files may be submitted by email, FTP, or on diskette. The DVI file(s) and PostScript files should be submitted only by FTP or on diskette unless they are encoded properly to submit through e-mail. (DVI files are binary and PostScript files tend to be very large.)

Files sent by electronic mail should be addressed to the Internet address pub-submit@ams.org. The subject line of the message should include the publication code to identify it as a Memoir. TeX source files, DVI files, and PostScript files can be transferred over the Internet by FTP to the Internet node e-math.ams.org (130.44.1.100).

Electronic graphics. Figures may be submitted to the AMS in an electronic format. The AMS recommends that graphics created electronically be saved in Encapsulated PostScript (EPS) format. This includes graphics originated via a graphics application as well as scanned photographs or other computer-generated images.

If the graphics package used does not support EPS output, the graphics file should be saved in one of the standard graphics formats—such as TIFF, PICT, GIF, etc.—rather than in an application-dependent format. Graphics files submitted in an application-dependent format are not likely to be used. No matter what method was used to produce the graphic, it is necessary to provide a paper copy to the AMS.

Authors using graphics packages for the creation of electronic art should also avoid the use of any lines thinner than 0.5 points in width. Many graphics packages allow the user to specify a "hairline" for a very thin line. Hairlines often look acceptable when proofed on a typical laser printer. However, when produced on a high-resolution laser imagesetter, hairlines become nearly invisible and will be lost entirely in the final printing process.

Screens should be set to values between 15% and 85%. Screens which fall outside of this range are too light or too dark to print correctly.

Any inquiries concerning a paper that has been accepted for publication should be sent directly to the Editorial Department, American Mathematical Society, P. O. Box 6248, Providence, RI 02940-6248.

Editors

This journal is designed particularly for long research papers (and groups of cognate papers) in pure and applied mathematics. Papers intended for publication in the *Memoirs* should be addressed to one of the following editors:

Ordinary differential equations, partial differential equations, and applied mathematics to JOHN MALLET-PARET, Division of Applied Mathematics, Brown University, Providence, RI 02912-9000; electronic mail: `jmp@cfm.brown.edu`.

Harmonic analysis, representation theory, and Lie theory to ROBERT J. STANTON, Department of Mathematics, The Ohio State University, 231 West 18th Avenue, Columbus, OH 43210-1174; electronic mail: `stanton@math.ohio-state.edu`.

Ergodic theory, dynamical systems, and abstract analysis to DANIEL J. RUDOLPH, Department of Mathematics, University of Maryland, College Park, MD 20742; e-mail: `djr@math.umd.edu`.

Real and harmonic analysis and geometric partial differential equations to WILLIAM BECKNER, Department of Mathematics, University of Texas at Austin, Austin, TX 78712; e-mail: `beckner@math.utexas.edu`.

Algebra and algebraic geometry to EFIM ZELMANOV, Department of Mathematics, Yale University, 10 Hillhouse Avenue, New Haven, CT 06520-8283; e-mail: `zelmanov@math.yale.edu`

Algebraic topology and cohomology of groups to STEWART PRIDDY, Department of Mathematics, Northwestern University, 2033 Sheridan Road, Evanston, IL 60208-2730; e-mail: `s_priddy@math.nwu.edu`.

Global analysis and differential geometry to CHUU-LIAN TERNG, Department of Mathematics, Northeastern University, Huntington Avenue, Boston, MA 02115-5096; e-mail: `terng@neu.edu`.

Probability and statistics to RODRIGO BAÑUELOS, Department of Mathematics, Purdue University, West Lafayette, IN 47907-1968; e-mail: `banuelos@math.purdue.edu`.

Combinatorics and Lie theory to PHILIP J. HANLON, Department of Mathematics, University of Michigan, Ann Arbor, MI 48109-1003; e-mail: `hanlon@math.lsa.umich.edu`.

Logic and universal algebra to THEODORE SLAMAN, Department of Mathematics, University of California at Berkeley, Berkeley, CA 94720; e-mail: `slaman@math.berkeley.edu`.

Number theory and arithmetic algebraic geometry to ALICE SILVERBERG, Department of Mathematics, Harvard University, 1 Oxford St.–Science Center, Cambridge, MA 02138; e-mail: `silver@math.ohio-state.edu`.

Complex analysis and complex geometry to DANIEL M. BURNS, Department of Mathematics, University of Michigan, Ann Arbor, MI 48109-1003; e-mail: `dburns@umich.edu`.

Algebraic geometry and commutative algebra to LAWRENCE EIN, Department of Mathematics, University of Illinois, 851 S. Morgan (M/C 249), Chicago, IL 60607-7045; email: `lawrence.man.ein@uic.edu`.

Geometric topology to JOHN LUECKE, Department of Mathematics, University of Texas at Austin, Austin, TX 78712; e-mail: `luecke@math.utexas.edu`.

All other communications to the editors should be addressed to the Managing Editor, PETER SHALEN, Department of Mathematics, University of Illinois, 851 S. Morgan (M/C 249), Chicago, IL 60607-7045; e-mail: `shalen@math.uic.edu`.

Selected Titles in This Series

(*Continued from the front of this publication*)

592 **P. Kirk and E. Klassen,** Analytic deformations of the spectrum of a family of Dirac operators on an odd-dimensional manifold with boundary, 1996

591 **Edward Cline, Brian Parshall, and Leonard Scott,** Stratifying endomorphism algebras, 1996

590 **Chris Jantzen,** Degenerate principal series for symplectic and odd-orthogonal groups, 1996

589 **James Damon,** Higher multiplicities and almost free divisors and complete intersections, 1996

588 **Dihua Jiang,** Degree 16 Standard L-function of $GSp(2) \times GSp(2)$, 1996

587 **Stéphane Jaffard and Yves Meyer,** Wavelet methods for pointwise regularity and local oscillations of functions, 1996

586 **Siegfried Echterhoff,** Crossed products with continuous trace, 1996

585 **Gilles Pisier,** The operator Hilbert space OH, complex interpolation and tensor norms, 1996

584 **Wayne W. Barrett, Charles R. Johnson, and Raphael Loewy,** The real positive definite completion problem: Cycle completability, 1996

583 **Jin Nakagawa,** Orders of a quartic field, 1996

582 **Darryl McCollough and Andy Miller,** Symmetric automorphisms of free products, 1996

581 **Martin U. Schmidt,** Integrable systems and Riemann surfaces of infinite genus, 1996

580 **Martin W. Liebeck and Gary M. Seitz,** Reductive subgroups of exceptional algebraic groups, 1996

579 **Samuel Kaplan,** Lebesgue theory in the bidual of $C(X)$, 1996

578 **Ale Jan Homburg,** Global aspects of homoclinic bifurcations of vector fields, 1996

577 **Freddy Dumortier and Robert Roussarie,** Canard cycles and center manifolds, 1996

576 **Grahame Bennett,** Factorizing the classical inequalities, 1996

575 **Dieter Heppel, Idun Reiten, and Sverre O. Smalø,** Tilting in Abelian categories and quasitilted algebras, 1996

574 **Michael Field,** Symmetry breaking for compact Lie groups, 1996

573 **Wayne Aitken,** An arithmetic Riemann-Roch theorem for singular arithmetic surfaces, 1996

572 **Ole H. Hald and Joyce R. McLaughlin,** Inverse nodal problems: Finding the potential from nodal lines, 1996

571 **Henry L. Kurland,** Intersection pairings on Conley indices, 1996

570 **Bernold Fiedler and Jürgen Scheurle,** Discretization of homoclinic orbits, rapid forcing and "invisible" chaos, 1996

569 **Eldar Straume,** Compact connected Lie transformation groups on spheres with low cohomogeneity, I, 1996

568 **Raúl E. Curto and Lawrence A. Fialkow,** Solution of the truncated complex moment problem for flat data, 1996

567 **Ran Levi,** On finite groups and homotopy theory, 1995

566 **Neil Robertson, Paul Seymour, and Robin Thomas,** Excluding infinite clique minors, 1995

565 **Huaxin Lin and N. Christopher Phillips,** Classification of direct limits of even Cuntz-circle algebras, 1995

564 **Wensheng Liu and Héctor J. Sussmann,** Shortest paths for sub-Riemannian metrics on rank-two distributions, 1995

563 **Fritz Gesztesy and Roman Svirsky,** (m)KdV solitons on the background of quasi-periodic finite-gap solutions, 1995

562 **John Lindsay Orr,** Triangular algebras and ideals of nest algebras, 1995

561 **Jane Gilman,** Two-generator discrete subgroups of $PSL(2, R)$, 1995

(See the AMS catalog for earlier titles)